REVOLUTIONARY WAR FORTS

CASEMATE | ILLUSTRATED | SPECIAL

⊆ CASEMATE | ILLUSTRATED | SPECIAL

REVOLUTIONARY WAR FORTS

New York

Michael Garlock

CISS0017

Published in the United States of America and Great Britain in 2023 by
CASEMATE PUBLISHERS
1950 Lawrence Road, Havertown, PA 19083, USA
and
The Old Music Hall, 106–108 Cowley Road, Oxford OX4 1JE, UK
Copyright © 2023 Michael Garlock

Hardback Edition: ISBN 978-1-63624-260-6
Digital Edition: ISBN 978-1-63624-261-3

A CIP record for this book is available from the British Library

Design by Battlefield Design
Printed and bound in the Czech Republic by FINIDR s.r.o.

For a complete list of Casemate titles, please contact:
CASEMATE PUBLISHERS (US)
Telephone (610) 853-9131
Fax (610) 853-9146
Email: casemate@casematepublishers.com
www.casematepublishers.com

CASEMATE PUBLISHERS (UK)
Telephone ((0)1226 734350
Email: casemate-uk@casematepublishers.co.uk
www.casematepublishers.co.uk

Title page image: French Castle at Fort Niagara. (Ad Meskens (CC BY-SA 3.0))

Contents page image: The Flag Bastion at Fort Ticonderoga, from the Detroit Publishing Company postcards series. (New York Public Library, The Miriam and Ira D. Wallach Division of Art, Prints and Photographs (CC0 1.0))

Contents page map: The frontier of New York in the Revolution. (*The Old New York Frontier* by Francis W. Halsey (1901))

Contents

Timeline

1625	The original iteration of Fort Amsterdam—aka Fort James, Fort Willem Hendrick (anglicized to Fort William Henry), Fort Anne, and Fort George—is built.
April 18, 1689	Boston revolt; Massachusetts Bay Colony reclaims government control.
1740	First basic iteration of Fort Herkimer is built.
May 28, 1754	French and Indian War starts, part of the wider Seven Years' War.
August 10, 1756	French troops under General Montcalm begin siege of the ultimately doomed Fort Oswego.
July 8, 1758	Battle of Carillon (or Ticonderoga), one of the bloodiest battles in the French and Indian War, proves a costly defeat for the British.

Fort Ticonderoga: Old Glory.
(Manuela Michailescu (CC BY-SA 4.0))

August 26, 1758–1762	British construction of Fort Stanwix to guard the Oneida Carry.
July 24, 1759	French relief force to Fort Niagara is ambushed at the battle of La Belle-Famille.
September 8, 1760	Pierre de Rigaud, Governor of New France, surrenders to British General Jeffrey Amherst.
February 10, 1763	French and Indian War concludes with a British victory (Treaty of Paris).
April 25, 1763–July 25, 1766	Pontiac's War signals dissatisfaction with British rule by the Native American confederation.
March 22, 1765	British Parliament enacts the Stamp Act. Violent demonstrations erupt in several colonies.
March 5, 1770	Boston Massacre.
December 16, 1773	Boston Tea Party.
April 19, 1775	Battles of Lexington and Concord, the siege of Boston.
May 10, 1775	Fort Ticonderoga captured by Ethan Allen, Benedict Arnold, and the Green Mountain Boys.
June 14, 1775	Congress approves creation of Continental Army; George Washington appointed commanding general.
December 31, 1775	Battle of Quebec; American attack repulsed by the British.
March 3–4, 1776	Battle of Nassau.
June 1776	Flagstaff Fort is rebuilt by the British on Signal Hill, Staten Island.
July 12, 1776	Fort Amsterdam's cannon engage British frigates; Fort Defiance's artillery engages HMS *Phoenix* and HMS *Rose*, to no effect.
August 27, 1776	Battle of Long Island aka battle of Brooklyn.
September 16, 1776	Battle of Harlem Heights.
October 11, 1776	Battle of Valcour Island.
October 29, 1776	Battle of White Plains.

November 16, 1776	Battle of Fort Washington; Howe claims a convincing victory.
November 20, 1776	Battle of Fort Lee and the fort falls to the British
December 26, 1776	Battle of Trenton.
May 23, 1777	Meigs Raid.
July 6, 1977	British place artillery on Mount Defiance; Americans abandon Fort Ticonderoga.
August 2–23, 1777	Siege of Fort Stanwix.
August 6, 1777	Battle of Oriskany.
August 22, 1777	Battle of Staten Island.
September 1, 1777	Siege of Fort Henry.
October 6, 1777	Battle of Forts Clinton and Montgomery.
September 19–October 7, 1777	Battles of Saratoga; surrender of British forces under General Burgoyne.
April 30, 1778	The Great Chain across the Hudson is completed.
July 16, 1779	Battle of Stony Point.
May 1780	British construct Fort Brooklyn in Brooklyn Heights.
July 11, 1780	Expédition Particulière; 5,000 French regulars arrive in North America.
September 23, 1780	Major John André is captured and Benedict Arnold's treason is exposed.
October 2, 1780	André is executed as a spy.
October 7, 1780	Battle of King's Mountain.
October 17, 1780	Sir John Johnson and the Mohawk firebrand Captain Joseph Brant attack Americans at Old Stone Fort.
October 19, 1780	Battle of Klock's Field, aka the battle of Failing's Orchard, or the battle of Nellis Flats, or the battle of Stone Arabia.
October 19, 1781	General Cornwallis and the British surrender at Yorktown.

Annual Fort Ward (Alexandria) Revolutionary War reenactment, 2008. (David from Washington, D.C., uploaded by Albert Herring (CC BY 2.0))

September 11–13, 1782	Siege of Fort Henry.
March 1783	British abandon Fort Golgotha, built out of gravestones by the ghoulish Lieutenant-Colonel Benjamin Thompson.
September 3, 1783	Treaty of Paris (1783) ends the American Revolutionary War.
November 25, 1783	Evacuation Day: British evacuate New York; Washington and the Continental Army return in triumph.
April 30, 1789	George Washington is inaugurated as the nation's first president.
1796	Six Northwest Territory forts and two Upstate New York forts under British control are ceded to the United States.

Revolutionary War Forts

QUEBEC

Kingston

Port Hope

TORONTO

LAKE ONTARIO

Ft Oswe

Ft Niagara

Niagara

Rochester

LAKE HURON

LAKE ERIE

Goshen

West Point

Mahopac

Bethel

New Haven

Guilford

Ft Montgomery

Ft Clinton

Peekskill

Ridgefield

Stepney

Milford

New Milford

Haverstraw

Bedford

Bridgeport

HUDSON R.

White Plains

Norwalk

LONG ISLAND SOUND

Ridgewood

Pompton

Ft Franklin

Port Jefferson

Oyster Bay

Ft Golgotha

Ft Salonga

Yaphank

Sunbury

Chatham Bridge

Ft Amsterdam

New York

LONG ISLAND

Ft Brooklyn

Westbury

Islip

Newark

Ft Defiance

Rahway

Ft Wadsworth / Flagstaff Fort

ATLANTIC OCEAN

HARRISBURG

Metuchen

RARITAN BAY

N

0 5 10 20
Miles

SUSQUEHA

Fort au Fer

Milltown

L. CHAMPLAIN

Vergennes

Ft Ticonderoga

Wentworth

Rutland

CONNECTICUT R.

Granville

Ft Stanwix

Ft Dayton

Herkimer

Utica

Ft Herkimer

Fort Klock

Fonda

Portsmouth

Winchester

NEW YORK

Old Stone Fort

Albany

Adams

MASSACHUSETTS

Boston

Springfield

Plymouth

Kingston

HUDSON R.

CONNECTICUT

Middletown

Falmouth

West Point

Ft Montgomery

Ft Decker

Ft Clinton

New Haven

DELAWARE R.

Wilkes-Barre

LONG ISLAND SOUND

Ft Franklin

Ft Salonga

Ft Golgotha

Ft Amsterdam

NEW YORK

PENNSYLVANIA

Ft Wadsworth / Flagstaff Fort

Ft Washington

Trenton

ATLANTIC OCEAN

Burlington

PHILADELPHIA

0   10   20   50
Miles

N

A 1628 woodcut by Matthaeus Merian published along with Theodore de Bry's earlier engravings in his 1628 book on the New World. The engraving shows the March 22, 1622 massacre when Powhatan Indians attacked Jamestown and outlying Virginia settlements. (Courtesy of Posterazzi)

| Introduction

Forts already existed when the Revolutionary War broke out. The genesis began in 1607 when the Jamestown settlers built their forts from which to defend themselves. These colonizers surrounded their villages with stockades (palisades) that were nothing more than rows of felled trees stuck vertically into the ground. Muskets were fired from loopholes (slits) in the stockade or from a blockhouse. There were also cookhouses, barracks for officers and regular soldiers, and storage for ammunition and weapons.

A facsimile of the stockade (palisade) at Jamestown Settlement that was established in 1607. (Ken Lund courtesy of James Fort Site, Historic Jamestowne, Colonial National Historical Park, Jamestown, Virginia (14239043919))

Ditches were sometimes employed—when filled with water they became moats—and castellated parapets were utilized that enabled defenders to fire their weapons from a barbette or protected platform. Wooden palisades were set at an angle on the front slope of the earth rampart to form a revetment. To create fields of fire, trees, bushes, and scrub were cleared from the area around the fort, and were used to make fascines or bundles of sticks tied together, or chandeliers that were X-shaped sawhorses connected by a bar that supported the fascines.

By and large many Revolutionary War forts were impermanent earthworks that had been constructed as a response to a threat. Many were constructed during the French and Indian War. Newly built forts took many forms. They included stockades, blockhouses, palisades, redoubts, rifle and artillery batteries, redans or V-shaped projections from a fortified line, fleches or detached V-shaped defensive works in an open field, camps, outposts, and garrisons.

Everything from a log cabin with loopholes for rifles to a stone castle that could accommodate hundreds of cannons was called a fort. The forts could be triangular, rectangular, square, pentagonal, hexagonal, or star-shaped. The star shape was an early favorite of the Americans as their French allies were innovative and instrumental in the construction of such forts. Often gabions or baskets and cages filled with rocks to build supports were used.

There were 18 Revolutionary War forts in New York alone. Some were large, others small. Men fought in them. They prayed and cursed, experienced triumph and tragedy, and lived and died. These are some of their stories.

Reenactors fire their cannon at the Jamestown Settlement, Virginia, c. 1615–25. (Mobilus In Mobili courtesy of Flickr)

Evolution of the forts

Many forts were simply open works with tall earth parapets on top of which guns could be brought to bear on attackers. Forts were armed with whatever iron or brass cannons could be acquired, regardless of caliber, age, or country of manufacture. All cannons worked the same way and it made no practical difference if the guns were Spanish, British, or American. The only differences were that brass is softer than iron and that fact had to be taken into consideration in terms of rate of fire and size of projectile. Occasionally, older guns would either misfire or even explode. Gun crew fatalities were not uncommon.

In addition to lending themselves to practical construction because the necessary material was usually if not always abundantly available, earthwork forts readily absorbed the impact and dispersed the energy caused by direct strikes from cannonballs and, obviously, bullets. Damage created by the impact of heavy projectiles could easily be repaired simply by adding more earth. Knowing they were relatively safe increased the morale of the defenders and in turn exponentially reinforced discipline up the chain of command. Moreover, only a modicum of technical expertise in the construction of forts was required in the building of earthwork forts.

Bastion comes from *trace italienne* or Italian outline. Bastion forts were typically polygons with bastions at the corners that eliminated dead zones and permitted fire from the curtain. Some bastion forts featured cavaliers which were elevated structures entirely inside the main fort itself. These forts saw development in the 15th and 16th centuries in response to the French invasion of Italy. Michelangelo used star forts in the defensive earthworks of Florence. For the next 300 years these forts proliferated throughout Europe. Vauban, Louis XIV's military engineer, took this particular form to its logical extreme.

In response to the advent of more powerful guns, walls were imbedded into ditches that were fronted by glacis or earth slopes that prevented them from being razed by hostile cannon fire. The defenders were relatively safe and could return fire with impunity.

The defense of Pisa, Italy, in 1500 against the Florentine and French armies was the first instance in which the new design proved its worth. Because original fortifications were becoming prone to succumbing to enemy bombardment, the Italians made an earth rampart at the most vulnerable part of their fort. In a second siege of the Venetian city of Padua in 1509, the fort was surrounded by a wide ditch that could be supported by flanking fire from gunports in projections that extended into the ditch. Star forts were also constructed by the Order of St. John in Malta in 1552.

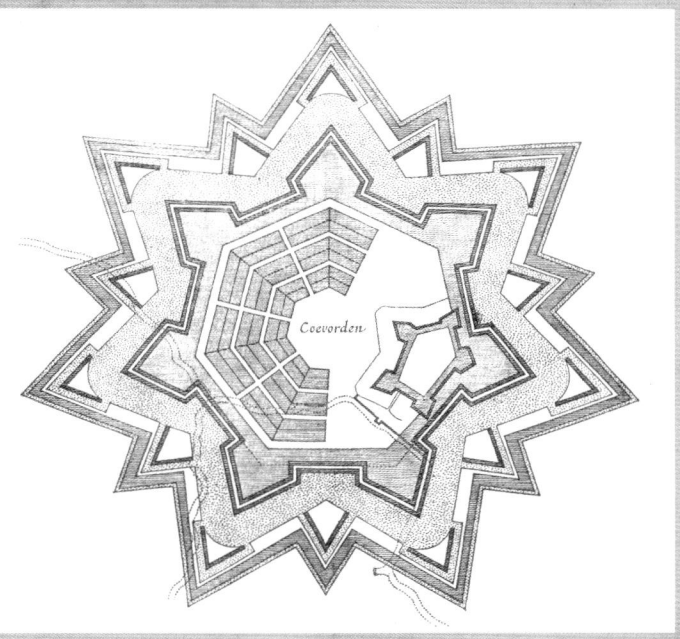

Plan of the fortress Coevorden in the Netherlands, 1647, a classic star fort with bastions, half-bastions, ravelins, and curtains. (Markus Schweiss (CC BY-SA 3.0) {{PD-US-expired}})

Military strategy obviously changed over the years. Cannons became more powerful and new attacking ideas and theories proliferated. The advent of the American Revolution gave many European engineers and designers who were out of work an opportunity to immigrate to America and ply their trade, which they did with great success.

Redraft of the Castello Plan, New Amsterdam, 1660. (J. W. Adams & I. N. Phelps Stokes courtesy of New-York Historical Society Library, Maps Collection (CC0 1.0) {{PD-US-expired}})

| Fort Amsterdam

Known variously throughout its long but somewhat mundane history as Fort James, Fort Willem Hendrick (anglicized to Fort William Henry), Fort Anne, and Fort George, this earthwork fort was constructed in 1625 on the southern tip of Manhattan at the confluence of the Hudson and East rivers. (The Hudson River was discovered by Henry Hudson, an Englishman who sailed for the Dutch East India Company. The name of his ship was the *Half Moon*.) The fort was designed by Cryn Fredericks, chief engineer of the New Netherlands colony. Construction was supervised by Willem Verhulst, second director of the New Amsterdam colony. The fort had four sides with a bastion on each corner and was built of hard-packed earth. Initial fortifications were constructed of whatever materials were found, either on site or relatively close by to enable expeditious construction. Timber and earth were the primary resources along with brushwood.

Later iterations of Fort Amsterdam were constructed of stone. It was located on a hill that sloped down to Bowling Green and Pearl Street and contained a church, barracks, a separate building for the director of the West India Company, and a spacious warehouse for the storage of valuable goods. Although protecting the New

Image of Fort Amsterdam on Manhattan, c. 1630. (Courtesy of New York Public Library (CC0 1.0))

Amsterdam colony from the French and British was the fort's primary mission, it was also the epicenter of trade and commerce. Soldiers garrisoned at the fort regularly drilled on what later became Whitehall Street. The fort was demolished in 1788 and the hill on which it stood was cut down and leveled.

In the autumn of 1664, Dutch settlers were taken aback when four British warships pulled into New Amsterdam harbor and demanded that the Dutch surrender. On September 8 Peter Stuyvesant prudently handed over the colony to the British without any needless loss of life. The fort was renamed Fort James in honor of James II of England and New Amsterdam was renamed New York in acknowledgement of James's title as the Duke of York.

The Dutch waited nine years before a fleet of 21 warships recaptured Manhattan in August 1673 in what was the Third Anglo-Dutch War. Again, the fort was renamed, this time Fort Willem Hendrick to honor William III, Prince of Orange, and New York was renamed New Orange. In 1674, the English again took control of the fort following the Treaty of Westminster which ended the war. The treaty also stipulated that there be established a diverse commission for the guidelines of trade, principally in the East Indies. It was signed on February 19, 1674 by Charles II of England and ratified by the States General of the Netherlands on March 5, 1674.

Once again the fort was renamed, reverting back to Fort James, while the area was again named New York. Thomas Dongan, 2nd Earl of Limerick, was appointed Vice-Admiral of the Navy and it was he who positioned gun batteries south of the fort. (The building of the fort denoted the sanctioned establishment of New York City as documented by its seal.)

After the "Glorious Revolution" (aka the English Revolution of 1688) that saw William and Mary ascend the English throne, Jacob Leisler, an upstart, wealthy German-born colonist who was a provincial militia captain, seized the fort in an action named Leisler's Rebellion. Representing the commoners against the wealthy leaders, chief among whom were Peter Stuyvesant and the Crown's representative in New York, Lieutenant-Governor Francis Nicholson, who promptly fled to England when the fort was seized, Leisler formed a government

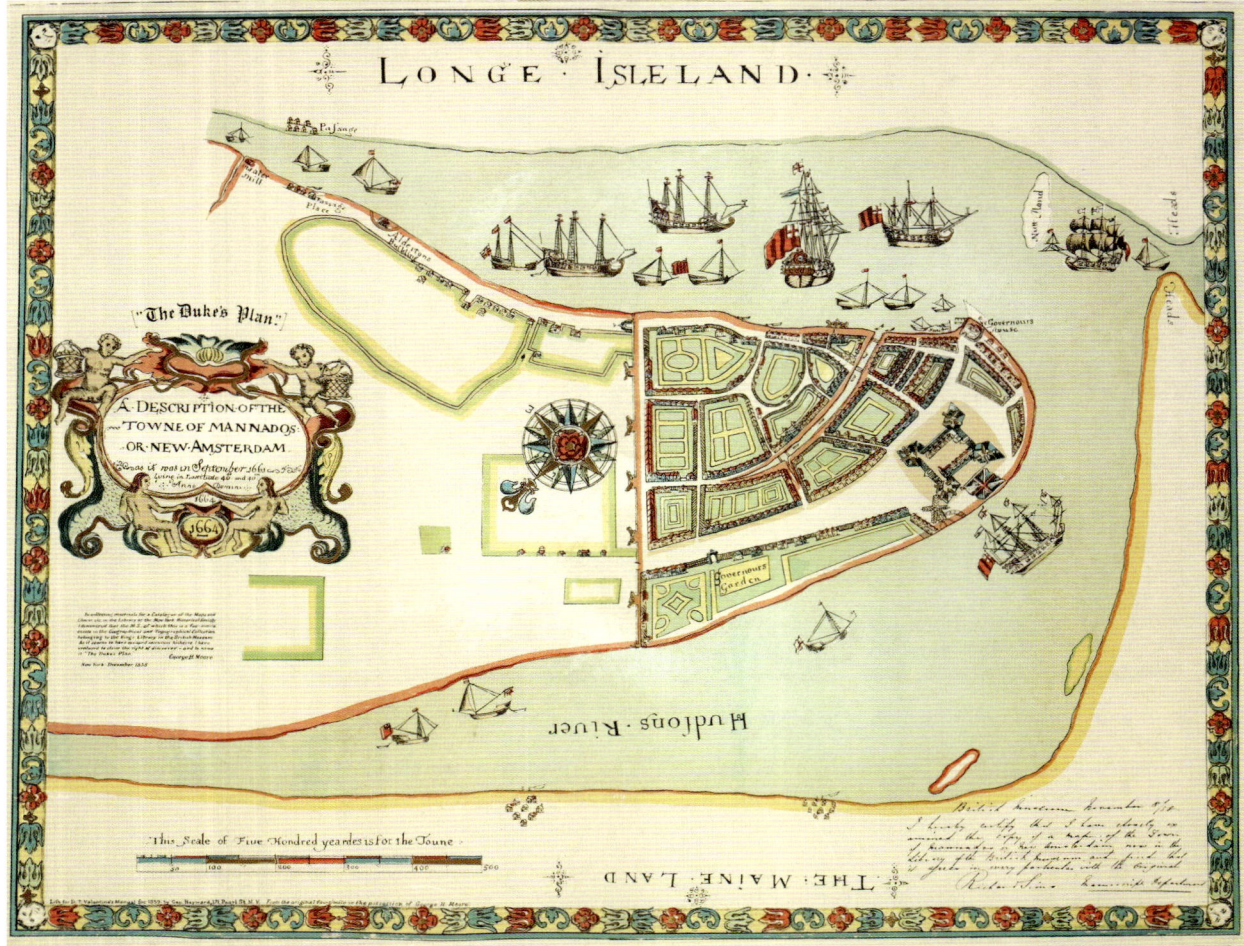

Longe Iseland or Map of Mannados or Manhattan, 1664, from The New York Public Library. {{PD-US-expired}}

of direct and popular representation. Oddly enough, he did not change the name of the fort. He was one of many colonists who vigorously resisted the united management that was called the Dominion of New England enforced by King James II on both New York and New England. Leisler's time as leader was short-lived and lasted only two years until 1690 when William's new governor finally arrived. It was he who renamed the fort Fort William and Henry to honor the new Protestant king of England.

The fort underwent a succession of name changes based on the English monarchs: Fort Anne for Queen Anne of Great Britain was followed by Fort George for the Georgian monarchs George I and George II of Great Britain, and George III of the United Kingdom. Perhaps, most notably, during the French and Indian War, 92 cannons were added to the battery in 1756.

After the British Parliament enacted the Stamp Act in 1765, which for the first time taxed a wide-ranging number of enterprises and business transactions, American rioters targeted the fort and spiked the battery's guns. The colonists under George Washington seized the fort in 1775 before the Revolutionary War actually began. In July 1776, the Declaration of Independence was read aloud in the common area of New York. General Washington and several brigades attended the ceremony. Afterwards a gilded lead equestrian statue of George III was torn town and the head was lopped off and melted down for musket balls.

Washington was informed on June 28 that the British fleet had set sail from Halifax on June 9, and was heading for New York. A day later, 45 British ships dropped anchor in Lower New York Bay. American troops hurried to their assigned posts. A week later, 130 British ships under the command of Admiral Richard Howe had anchored off Staten Island and, on July 2, British troops began landing on Staten Island. The Americans stationed there did the prudent thing and fled.

On July 12, the British ships *Phoenix* and *Rose* sailed out the harbor and made for the mouth of the Hudson River. They were met with fire from the guns of Fort George, Fort Defiance, and Governors Island. Undeterred, the British lobbed shells into New York City. Sailing past Fort Washington before heading up the Hudson, the ships arrived at Tarrytown where they hoped to cut off supplies from New England and to inspire Loyalist support. Six Americans were killed when their cannon blew up.

On August 27 the fight began in earnest.

The Americans were led by Commander-in-Chief General George Washington who had approximately 10,000 soldiers facing the British, led by General William Howe, who had an army of 20,000. He was aided by his brother, Admiral Richard Howe, who commanded the massive fleet of more than 400 vessels including 73 warships. The total British force was 32,000. Other British commanders were Charles Cornwallis, Henry Clinton, William Erskine, James Grant, Charles Mawhood, and Francis Smith.

The admiral's brother, General Sir William Howe, 5th Viscount Howe. (Richard Purcell aka Charles Corbutt courtesy of ColorMezzotint (CC0 1.0) {{PD-US-expired}})

Admiral Rich Howe (1726–99). (Courtesy of Royal Museums Greenwich)

After securing Boston from the British, Washington presciently realized that New York City would be the next to be attacked. The city was and still is surrounded by water. Because he didn't have any warships, Washington had moved his troops to the city in April 1776. Three divisions were stationed at the southern end of Manhattan, one division was positioned in northern Manhattan, and one division on Long Island. Other American commanders included William Alexander, Thomas Mifflin, Henry Knox, and John Sullivan. Major-General Nathanael Greene was originally supposed to command the troops on Long Island, but he became ill so General John Sullivan took over as commander. On August 24 Sullivan was replaced by one of the more colorful characters of the war, Israel Putnam. His role was to direct the defenses from Brooklyn Heights. The British stationed their troops on nearby Staten Island which was sparsely populated at the time.

The British wore red coats with bearskin caps for the grenadiers, tricorne hats for the battalion companies, and caps for the light infantry. Their Hessian allies wore blue coats while the Hessian grenadiers wore the Prussian-style brass-fronted miter cap. The Americans had no uniforms and wore what were basically street clothes. The British 17th Light Dragoons and desultory mounted American units were the only cavalry in the ensuing battle. The British infantry was composed of amalgamated battalions of light infantry, grenadiers, and foot guards, including the 1st, 2nd, and 3rd Guards, the 15th, 22nd, 28th, 33rd, 35th, 37th, 38th, 42nd (Black Watch), 43rd, 44th, 45th, 49th, and 63rd Regiments of Foot, and Fraser's Highlanders.

Prior to the battle, British commanders Clinton and Cornwallis and 4,000 troops landed at Gravesend Bay, at 8 a.m. on August 22. Four hours later their force had increased to 15,000 with 40 artillery pieces.

Henry Knox, the American artillery commander, convinced Washington to assign 400 to 500 soldiers who did not have rifles to serve as artillery gun crews. At the northern end of Manhattan General Nathanael Greene decided to construct Fort Washington with Fort Constitution (later named Fort Lee) to be situated opposite Fort Washington on the Hudson River. These forts were meant to discourage or prevent British ships from sailing up the Hudson. Unlike the two- and three-casement, mortar and brick forts that predominated during the Civil War with their gigantic smooth-bore cannons, these Revolutionary War forts were relatively easy and quick to construct.

On August 22, 4,000 British regulars departed Staten Island under the joint command of Clinton and Cornwallis and landed on Long Island. Pennsylvanian riflemen commanded by Colonel Edward Hand did not oppose the landing and prudently withdrew. Cornwallis pushed farther and established a camp in what is now Flatbush. His orders were to stay where he was. General Howe then laid siege to the Americans there, digging trenches and establishing lines of circumvallation around the American positions.

It was obvious that the Americans were outmanned by the numerically superior British, outfought by better-trained and more disciplined regular troops, and outgunned by Howe's superior firepower. Washington and his ragtag army found themselves in the unenviable and untenable position of being surrounded on Brooklyn Heights, with the East River to their backs. Despite this, 1,200 additional troops from Manhattan were ordered to Brooklyn: two Pennsylvanian regiments commanded by Thomas Mifflin, plus Colonel John Glover's regiment from Marblehead, Massachusetts. On the afternoon of August 28, there were numerous skirmishes while rain started to fall. American guns bombarded the British ships. At 11 p.m. Glover and his men, most of whom were fishermen and sailors, began to evacuate the sick and wounded. It was Mifflin's turn to evacuate at 4 a.m. (on August 29) but he returned to his defensive position when it was learned that the order had been mistaken. A fog fortuitously rolled in at daybreak, concealing the final stages of the evacuation. By 7 a.m. the entire American force, totaling 9,000 soldiers, had been evacuated without a single loss of life. General Washington reputedly rode in the last boat. The Americans had lived to fight another day.

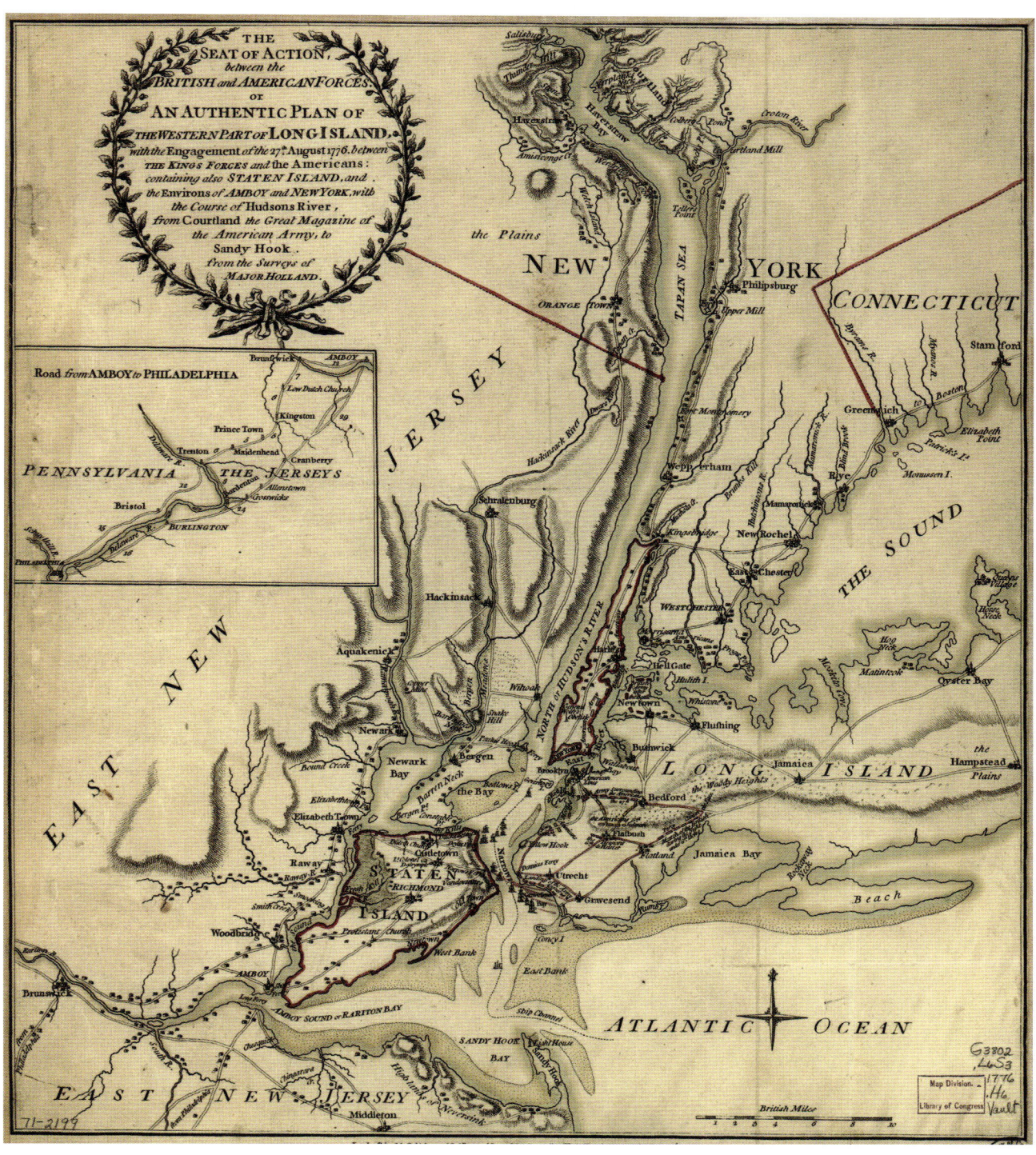

The seat of action, between the British and American forces; an authentic plan of the western part of Long Island, with the engagement of August 27, 1776, between the King's forces and the Americans: containing also Staten Island, and the environs of Amboy and New York, with the course of Hudson's River from Courtland, the great magazine of the American Army, to Sandy Hook. (S. Holland, R. Sayer & John Bennett (firm) courtesy of Library of Congress (CC0 1.0))

The Battle of Long Island by Domenick D'Andrea for the state of Delaware and Maryland, 1776. The Delaware Regiment in action. (Courtesy of Flickr (CC0 1.0) {{PD-US-expired}})

Following the July 12 naval exchange, the guns of Fort Amsterdam again engaged British frigates and exchanged volleys with British guns on Governors Island between September 2 and 14. The distance from the tip of Manhattan to Governors Island is 6.9 miles (11.1 kilometers). Obviously, firing at a moving ship from a stationary position presented the fort's gunners with challenges. Their targets were moving from side to side, forward and backward, and bobbing up and down in the water. The agile British frigates also struggled to return fire accurate due to the ships' constant movement, however the walls of a fort were much larger than a ship and presumably easier to hit.

This was the only action the fort saw, and its guns were never fired in anger again. Because no casualties for either the Americans or the British are listed in historical records it is reasonable to assume that either there weren't any, or if there were they simply weren't recorded. In September 1776, the British captured the fort and lower Manhattan and subsequently held New York for the entire war. British casualties in the battle of Long Island—also known as the battle of Brooklyn or Brooklyn Heights—were 64 killed, 293 wounded, and 31 missing. The Americans suffered 2,179 killed, wounded, or captured. The American commander John Sullivan was captured.

Commodore Walker's Action: The Privateer Boscawen Engaging a Fleet of French Ships, 23rd May 1745 by Charles Brooking. The *Boscawen* (center, flying the Royal Navy red ensign) was formerly the French frigate *Médée* that had been captured by the British. (Courtesy of Royal Museums Greenwich)

The frigates

From the Italian *fregata*; the Dutch *fregat*; the Spanish/Catalan/Portuguese/Sicilian *fregata*, and the French: *fregate*), frigates of the time were full-rigged ships, meaning they were square rigged on all three masts, built for speed thanks to a long hull and maneuverability, and were more lightly armed than a ship of the line. They had at least 28 guns on the continuous upper deck as opposed to ships of the line that had two continuous decks that carried guns.

The Royal Navy copied the new French frigates, namely the *Médée*, during the War of Austrian Succession (1740–48), primarily because the French ships had superior handling closer inshore. The average hull length was 135 feet (41 meters), with an average draft of 13 feet (4 meters); speed under full sail with a favorable wind usually topped out at 14 knots (16 mph, 26 kph). Capable of carrying six months' worth of provisions, these hard-working vessels often acted independently of a larger fleet or armada. A posting to a frigate was desirable because they frequently saw action in attacking smugglers and privateers, which offered increased chances for promotion and prize money.

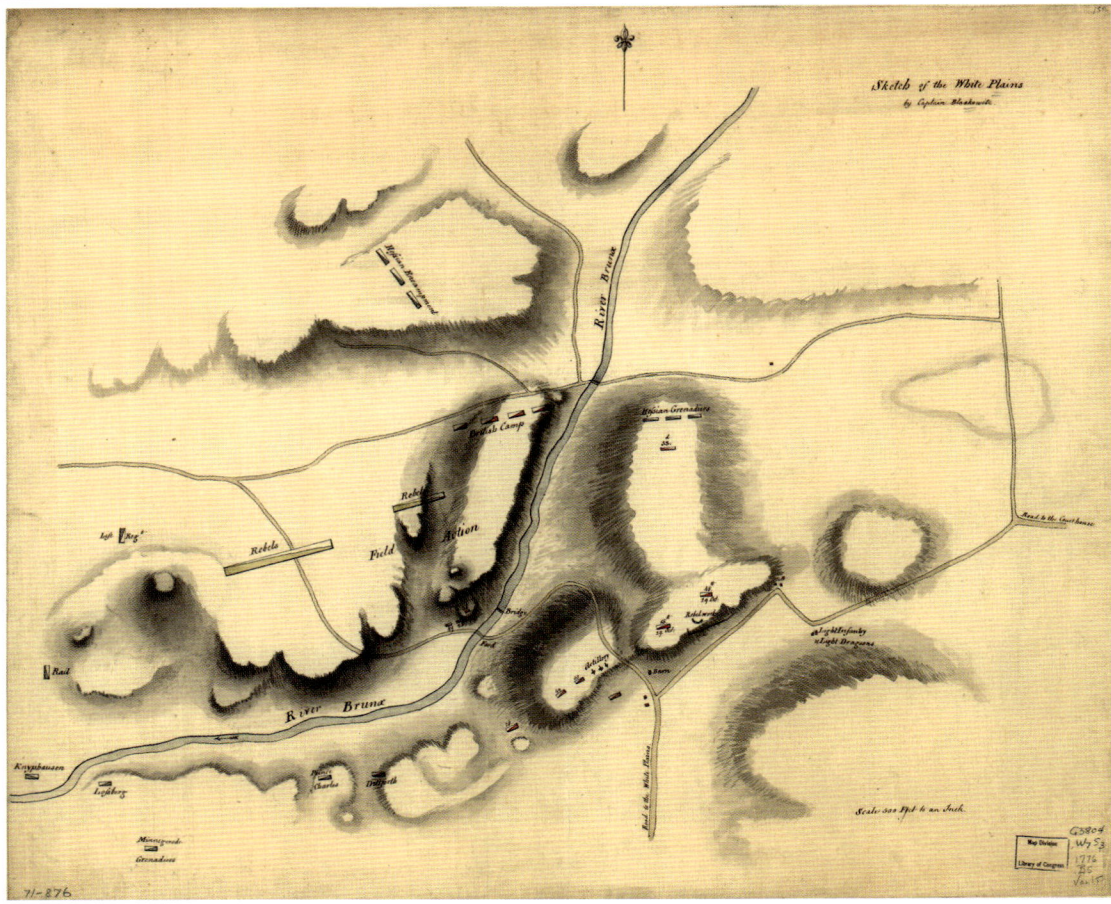

Sketch of the White Plains. (Charles Blaskowitz courtesy of the Library of Congress (CC0 1.0))

Subsequently, Howe landed a considerable force at Kis Bay and occupied it. The Americans counterattacked in the middle of September at Harlem Heights, but were defeated at White Plains and again at Fort Washington, prompting General Washington and his army to make a strategic withdrawal across New Jersey and into Pennsylvania. On September 22 a suspicious fire destroyed over a quarter of New York City. Nathan Hale was executed by the British for spying.

The captured Americans suffered a grisly fate; fewer than half survived. They were first kept on prison ships in Wallabout Bay before being transferred to other sites such as the Middle Dutch Church. Starved and denied proper medical attention, many succumbed to smallpox. Up to 256 soldiers belonging to Colonel William Smallwood's First Maryland Regiment were buried in a mass grave, the exact location of which has never been discovered.

The American fort had done its duty. Eager to aid their fellow countrymen, the defenders had fired their guns with enthusiasm and malicious intent. Other forts were also going to play a pivotal role in the Revolutionary War.

On November 25, 1783, Evacuation Day, the Americans took control of the fort after the British left. Five years later the fort was razed.

The fort today:

The site was later redeveloped for the Alexander Hamilton U.S. Customs House that was first located at 55 Water Street.

Fort Brooklyn

This British star fort, i.e., a fort with projecting angles that literally resembles a star, was constructed in May 1780 to support the occupation of Brooklyn. It was located in Brooklyn Heights near what are now Henry and Pierrepont Streets, around four blocks from Fort Sterling. Surrounded by an encircling ditch, the fort's ramparts were 40–50 feet (12.1–16.2 meters) above the ditch. The fort was 450 square feet (42 square meters) in extent. Each angle had a bastion, and the fort included a barracks and two magazines. After the British had left, the fort was demolished between 1823 and 1825 for development.

Fort Brooklyn stood just south of where Fort Sterling is shown on this section of a plan of the battle of Brooklyn, 1776. (Stiles, S. E., *History of the City of Brooklyn, Vol 1*, page 250)

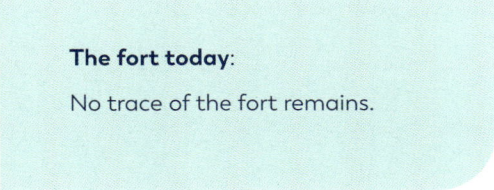

The fort today:

No trace of the fort remains.

Arming the frigates

The early British frigates were armed with 28 guns that included an upper-deck battery of 24 x 9-pounder guns. The other four, smaller-caliber guns were on the quarterdeck. These were followed by frigates with 26 x 12-pounder guns and six to 10 smaller guns on the quarterdeck and forecastle. The 9-pounder carriage-mounted, breech-loading, smooth-bore guns were manufactured by Woolwich Arsenal. The common shell for these guns weighed 9.1 pounds (4.1 kilograms), while shrapnel shells were slightly heavier, weighing 9.8 pounds (4.4 kilograms). Their muzzle velocity was 1,330 feet per second (405 meters per second).

A typical carriage-mounted 9-pounder cannon found on British frigates. (Dave Pape)

The 6-pounder carriage weighed 900 pounds (408 kilograms). The weight of the gun barrel was 785 pounds (356.1 kilograms) and the diameter of the bore was 4.62 inches (11.73 centimeters). After firing, the gun would recoil and move the cannon backward, adding to the time it took the 10-man crews to reposition, load, and fire. A well-trained crew could fire a round every 20 seconds. The guns fired a variety of projectiles such as solid shot, explosive-filled balls, grapeshot, canister shot, chain shot and shrapnel shot. Chain shot entailed two halves of a cannonball joined together by a chain or metal bar. After the gun was fired, the two halves would separate and either tumble or spin. They were notoriously inaccurate, and used against infantry, masts, and rigging.

Fort Clinton

The Hudson River is 315 miles (507 kilometers) long. Its maximum elevation is 1,770 feet (540 meters), its maximum depth is 202 feet (62 meters), and its average depth is 30 feet (9.1 meters). It is fed by eight tributaries on its left bank and nine on its right bank. It was called *Ka non:no* or *Ca-ho-ha-ta-te-a* by the Haudenosaunee, or *Mahicannittuk* by the Mohicans who inhabited lower portions of the river. The source is Lake Tear of the Clouds in the Adirondack Park. There the river has an elevation of 4,322 feet (1,317 meters). When the river reaches Calamity Brook, it is cartographically known as the Hudson.

Two forts were constructed in 1776: Fort Clinton (not to be confused with Fort or Castle Clinton in Battery Park, New York City, or Fort Clinton at West Point) and its sister fort, Fort Montgomery. Fort Clinton overlapped the convergence of Popolopen Creek on the southern side of Popolopen Gorge with Fort Montgomery to the north.

A plan of the Forts Montgomery and Clinton as surveyed by Major Samuel Holland, His Majesty's forces under the command of General Sir Henry Clinton, October 7, 1777. (Courtesy of Library of Congress (CC BY-NC-SA))

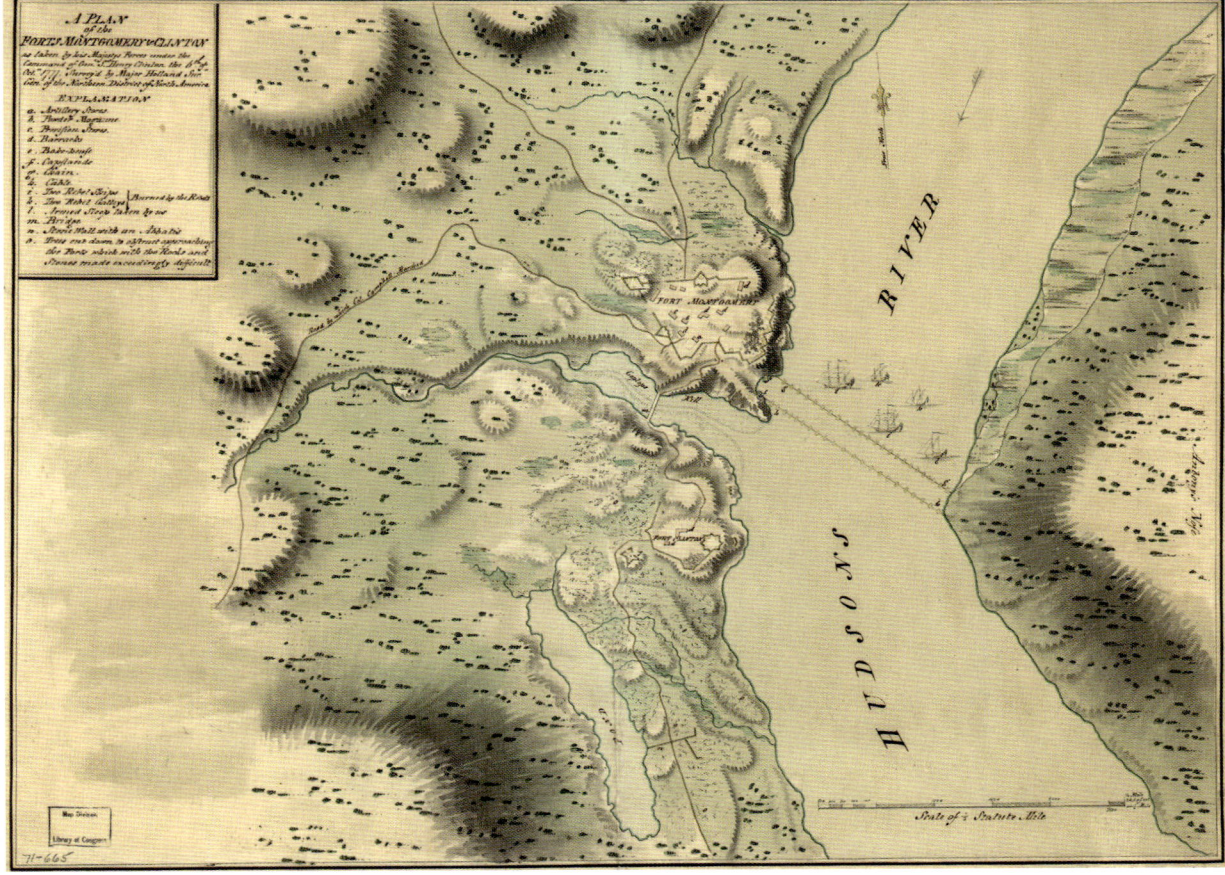

Popolopen Creek, also known as Popolopen Brook, is fed by Weyant's Pond, Stillwell Lake, Mine Lake, and Lake Popolopen, and is approximately 10.4 miles (16.7 kilometers) long.

The British knew that if they managed to control the upper reaches of the Hudson, this in turn would allow them free access to the lower part of the river, all the way to New York, throttling trade and the ability of the Americans to use the waterway as a means of introducing their own ships into any future clashes. Moreover, the Hudson River valley was a critical area because it allowed the British to move troops and supplies between the New England states and states farther south.

Moreover, the Hudson's proximity to Lake George and Lake Champlain meant that if the Royal Navy could control the river, it would in turn allow them to control the entire route from Montréal to New York City. The Hudson runs approximately parallel to these two lakes and is very roughly 100 miles (160 kilometers) from each. Both lakes are large with long shorelines, and lend themselves to both vessel transport and fort construction. Lake George is 32 miles (51 kilometers) long by 1–3 miles (1.6–5 kilometers) wide, with an average depth of 200 feet (60 meters). Its surface area is 28,200 acres (11,412 hectares) or 44 square miles (114 square kilometers) and it has an elevation of 322 feet (9 meters).

An ingenious American solution was to stretch an enormous wrought-iron chain across the Hudson, just below the surface, from Fort Montgomery to Anthony's Nose on the east bank. (Anthony's Nose, on a grant patent issued to Pierre Van Cortlandt because he owned the mountain, is a 910 feet- (277 meter-) high peak and was reputedly named after a pre-Revolutionary War sea captain, Anthony Hogan, whose proboscis was said to resemble Cyrano de Bergerac's. A chain would do much to secure the waterway for the Americans, and to prevent marauding, shallow-draft British frigates from attacking other installations along the river and generally wreaking havoc. Near Fort Montgomery the Hudson River is 422 feet wide (129 meters). No English captain would dare risk damaging his ship's hull and possibly causing the vessel to sink by running into the chain below the water surface.

Fort Clinton was erected on ground higher than its sister fort, Fort Montgomery, but its garrison consisted of only 300 soldiers who no doubt took solace in the fact that its defenses consisting of smooth-bore cannons were more complete.

A battle was in the offing as the British prepared to attack. It would come to be known as the battle of Forts Clinton and Montgomery.

The British were commanded by Sir Henry Clinton (no relation to the American commanders); Sir James Wallace (who in July 1776 became the captain of the 50-gun ship HMS *Experiment*); Lieutenant-General Sir James Vaughan, who had been wounded in his

Links of the great chain that spanned the Hudson. (Gevixel (CC BY-SA 4.0))

British commander General Sir Henry Clinton. From a painting by Andrea Soldi. (Courtesy of the American Museum in Britain (CC0 1.0) {{PD-US-expired}})

American commander General George Clinton by Ezra Ames. (Courtesy of www. nyhistory.org (CC0 1.0) {{PD-US-expired}})

thigh while leading the grenadiers at the battle of Long Island (he was also to lead a column during the assaults on forts Clinton and Montgomery where his horse was killed under him); and, lastly, Edmund Fanning, a colonial administrator. The British force of over 2,000 troops consisted of a detachment of the 17th Regiment of Light Dragoons, the 7th Regiment of Foot Royal Fusiliers, 17th Regiment of Foot, 26th Regiment of Foot, 26th Regiment of Foot, 52nd Regiment of Foot, 57th Regiment of Foot, 63rd Regiment of Foot, and a company of the 1st Battalion 71st Regiment of Foot. They were augmented by German troops in the form of a grenadier company, 1st Anspach-Beyreuth Regiment, and Regiment von Trumbach (Landgraviate of Hesse-Kassal).

The Americans were led by George Clinton and James Clinton (the pugilistic Israel Putnam and his soldiers were held in reserve at Peekskill). The 600-strong American force was composed of the Loyal American Regiment, Emmerich's Chasseurs, New York Volunteers, King's American Regiment, and the King's Orange Rangers.

It was foggy on the morning of October 6, 1777, when Sir Henry Clinton and slightly more than 2,100 soldiers landed at Stony Point on the west side of the Hudson. He then divided his force into two, and marched on the American forts. Lieutenant-Colonel Campbell led 900 men, 52nd and 57th Regiments, some Hessian chasseurs, and approximately 400 Loyalists under Beverly Robinson on a seven-mile (11 kilometers) slog through a gorge toward Fort Montgomery, while Sir Henry Clinton waited at a place called Doodletown before marching on Fort Clinton. Campbell positioned the British regiments on his right flank, the Germans in the center, and the Loyalists on the left flank.

There were only about 100 men defending Fort Montgomery. Captain John Lamb had established a defensive position a mile from the fort and engaged the advancing British with a small-caliber gun, probably an 8-pounder. Taking fire, the Americans retreated further after spiking their gun and made one last stand closer to the fort, this time firing a larger gun, a 12-pounder.

American commander General James Clinton by Charles Balthazar Julien Fevret de Saint-Mémin. (National Portrait Gallery, Smithsonian Institution; gift of Mr. and Mrs. Paul Mellon (CC0 1.0))

A short while later, Fort Clinton, also defended by abatis (rows of branches with sharpened points) was attacked by the 63rd Foot from the northwest, while the 7th and 26th Regiments and a company of Anspach grenadiers attacked from the west. This was followed by a detachment of 17th Light Dragoons, the 26th Foot, and the remaining German and British companies delivering a sledgehammer blow to the fort.

Both forts were captured after about an hour. General James Clinton, now wounded, led his surviving men though Popolopen Gorge and escaped to gunboats waiting to take them across the river to safety. The British lost 41 men killed in action with 142 wounded, while the Americans lost 75 killed and approximately 263 captured, most from Fort Clinton.

The fort today:

Like many other forts, Fort Clinton succumbed to the ever-increasing demands of development and was demolished to make way for a state road and a bridge. Scattered remnants are to be found within the confines of Bear Mountain State Park. In 1972 it was designated as a National Historic Landmark and was also placed on the National Register of Historic Places.

Ship rating

A first-, second-, or third-rate ship was a ship of the line. The first rates were reserved for admirals who commanded a fleet and had enough room to accommodate the admiral's staff. The second rate was a less expensive alternative to a first rate. The early first- and second-rate ships had three decks: the lower, middle, and upper decks. The largest third-rate ship had 80 cannons in addition to smaller guns on the quarterdeck, forecastle, and poop deck. From the 1750s these vessels were constructed with only two continuous decks (lower and upper) while smaller weapons were still positioned on the quarterdeck, forecastle, and the poop deck if the ship had one.

After January 1812 carronades were added to the ship's already impressive array of firepower and were included in the number of guns the ships carried. The carronade had a short range and a short barrel and was often mounted on a slide. Some ships might be armed with 12-, 18-, 24-, or 32-pounder carronades. Many of these ships were three-deckers and had more than 100 guns. The new iteration of second-rate ships had two continuous decks and carried 80 or more guns. Third-rate ships had two decks and less than 80 cannons.

A sketch by Pharamond Blanchard of a carronade aboard the French frigate *Médée*. (Courtesy of Service historique de la Défense)

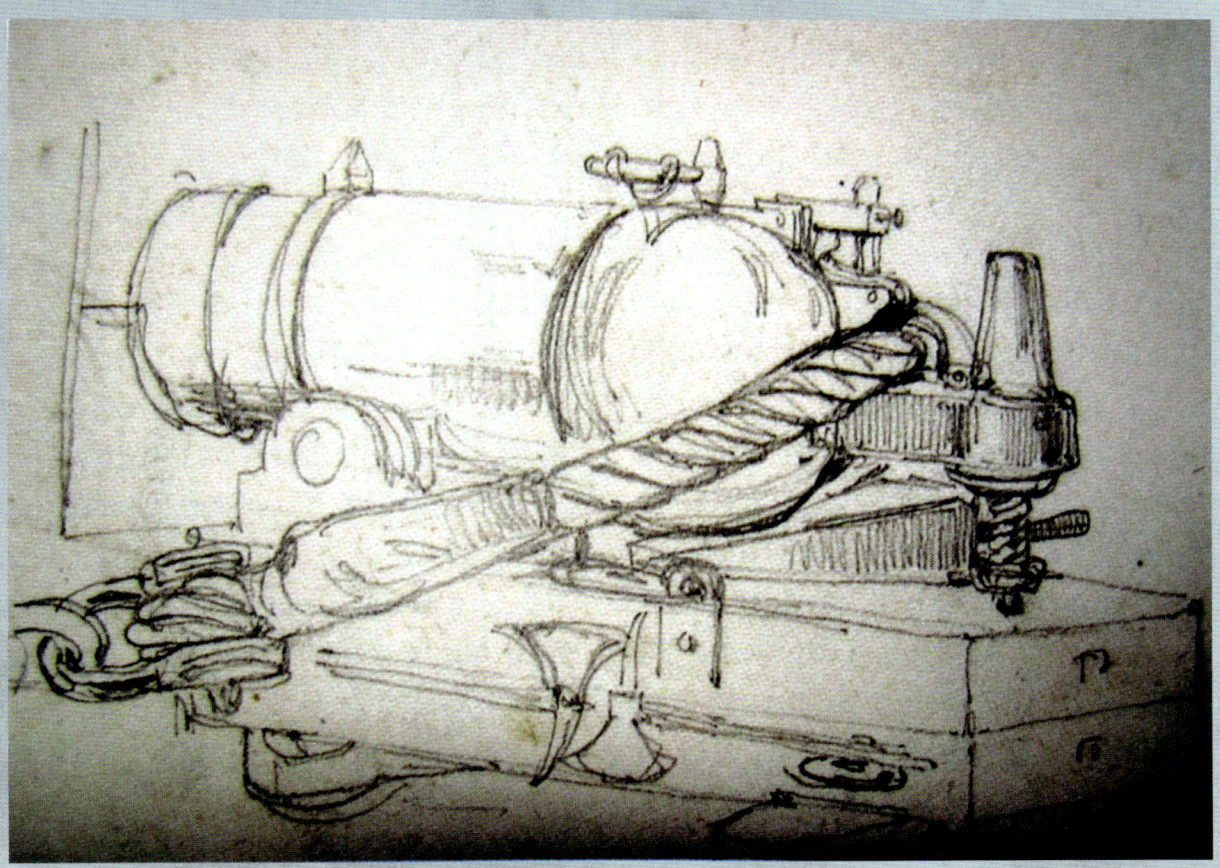

A 64-gun French ship of the line c. 1750–70 (author unknown). (Courtesy of Musée national de la Marine)

Until 1756, the smaller fourth-rate ships that had between 50 and 60 guns on two decks were considered ships of the line, but it was determined that these vessels weren't big enough to successfully participate in pitched battles. However, the larger fourth-rate ships with 60 guns were still regarded as ships of the line. The 60-gun fourth rate was succeeded by the 64-gun third rate over the following decades. Fourth-rate ships were used for escorting convoys or as flagships on distant stations.

The difference between a fourth rate and a fifth rate was often indistinguishable. From 1690 an upper-end fifth rate could and often did include both two-deckers that carried between 40 and 44 cannons and what was known as demi-batterie i.e., ships whose larger and heavier guns were located on the lower deck where the extra space was utilized for row ports and a full, numerically larger number of smaller and less powerful guns on the upper deck. These were gradually phased out when it became apparent that in rough weather it was impossible to open the lower-deck gunports.

Ships of the line never included fifth- and sixth-rate ships but by the middle of the 18th century a new fifth-rate ship was introduced, known as the classic frigate. It had no gunports on the lower deck. The main battery was on the upper deck which meant all the guns could be fired no matter how inclement the weather. Frigates played an important role in fights that occurred on rivers and harbors during the American Revolution and were a valuable tactical asset of the Royal Navy. Frigates could be employed as squadrons to fight an enemy force that did not have ships of the line.

Fort Clinton at West Point

This brick and masonry fort, now part of the United States Military Academy, was constructed between 1778 and 1790. The first builder was Captain Louis de la Radière and the fort was completed by Colonel Andrzej Tadeusz Bonawentura Kościuszko (Polish spelling; in English, Andrew Thaddeus Bonaventure Kościuszko), a Polish military engineer, statesman, and national hero in Poland, Lithuania, Belarus, and the United States. The fort was originally named Fort Arnold after Benedict Arnold but after Arnold's trial for spying and his subsequent hanging, it was understandably changed to Fort Clinton.

American military commanders had knowledge of Vauban's techniques, specifically as applied to siege warfare. His system involved digging a trench three to four feet deep, parallel to the protective guns of a city or fort but beyond the range of the city's or fort's cannons or rifles. The earth from the trench was used to create a parapet. Then the attackers dug a second trench diagonally towards the city or fort in a zigzag fashion until the attacker's artillery could reach the city or fort. A second parallel trench was dug, artillery was moved into this trench, and the assault was begun.

Fort Clinton, West Point. (Courtesy Library of Congress (CC0 1.0) {{PD-US-expired}})

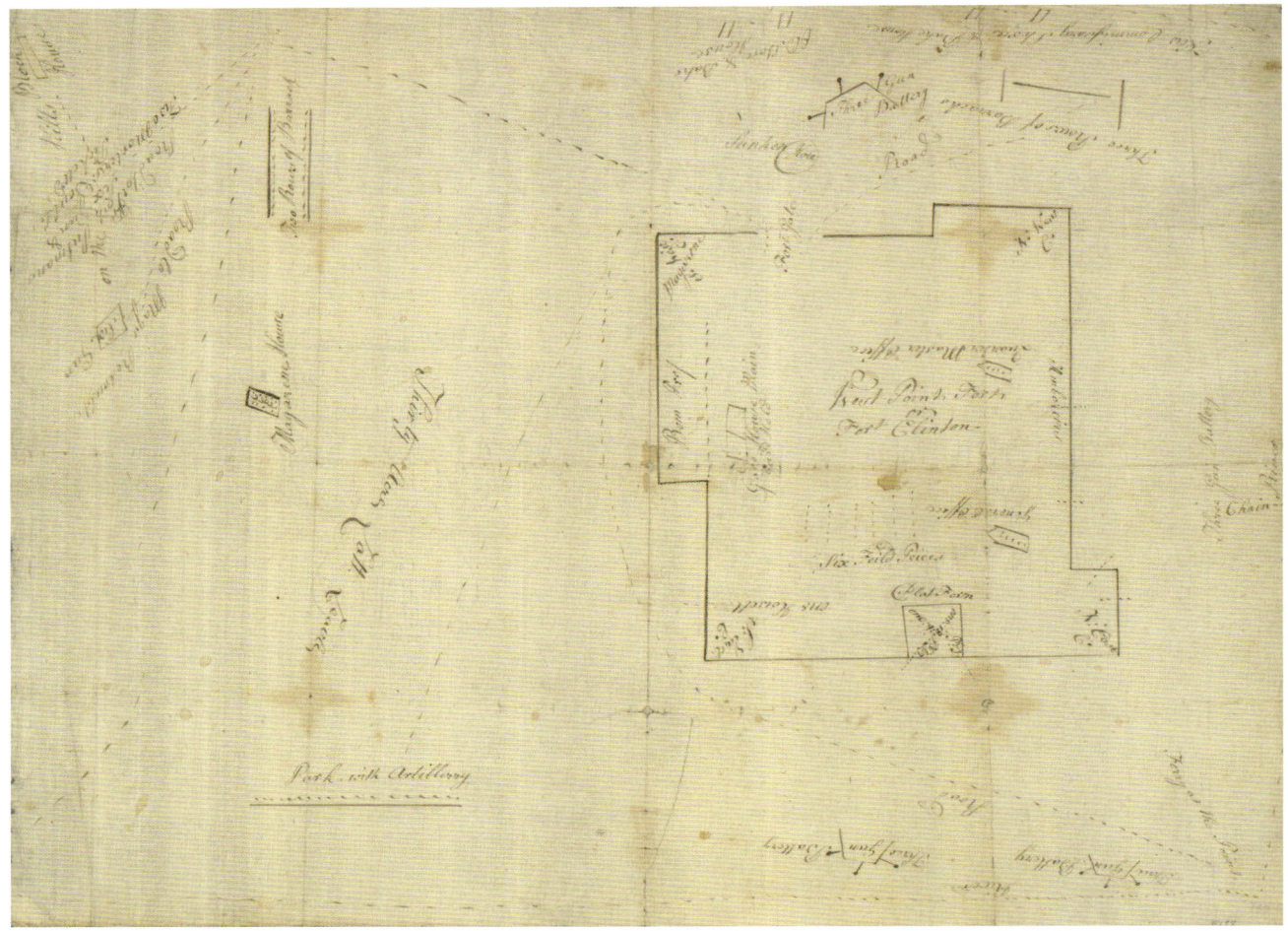

Pen and ink, outline map of the fort showing the positions of its guns and its relationship to other defenses and structures of the West Point defenses. (Courtesy of William L. Clements Library (CC0 1.0) {{PD-US-expired}})

While in Paris, the enterprising Kościuszko learned that American colonists needed civil engineers. In June 1776, he took a ship across the Atlantic, but the vessel was shipwrecked off Martinique. Undeterred, he finally made it to Philadelphia in August. Two months later, in October, John Hancock made him a colonel in the Continental Army and Benjamin Franklin employed him to design and build forts on the Delaware River. In May 1777, he found himself in New York evaluating Fort Ticonderoga's defenses. His advice was to strengthen an adjacent hill that overlooked the fort with cannons.

His superiors disagreed, contending that it wasn't feasible to move heavy artillery pieces (probably 32-pounders) up such a steep slope. The British, led by General John Burgoyne, thought otherwise; in July, the general slogged down from Canada with a force of 8,000 well-trained men and six heavy guns up to the top of the hill where they promptly bombarded the fort. The Americans had no choice but to evacuate. Kościuszko then designed a floating log bridge that greatly facilitated the American withdrawal with a minimum loss of life.

Portrait of M. Sebastien Le Prestre de Vauban by Louis Bernard after François de Troy. (Courtesy National Gallery of Art (CC0 1.0))

Sebastien Le Prestre de Vauban Apart from being the preeminent world leader in fort design and siege strategy, Vauban also designed the socket bayonet that attached to the outside of a soldier's firearm, and devised the innovative tactic of ricochet gunfire. He favored the French Army using the new flintlock musket that would replace the less efficient matchlock musket. Together, these two adaptions influenced the decline of what was called the age of the Spanish tercio that saw massed infantry armed with a combination of muskets and pikes. In turn, this allowed commanders to organize their infantry in thinner formations, thus revolutionizing tactics. Vauban also built permanent barracks for French soldiers and created the first professionally proficient force of military engineers. The designs of Fort Ticonderoga and Fort William Henry in New York, and Fort Monroe in Virginia, are just some of the many examples of the influence Vauban had on fort design.

Tadeusz Kościuszko was best known in his native Poland for leading the uprising in 1794 against foreign rule by Russia and Prussia. However, his journey to America and the role he played in designing some of the Revolutionary War's most iconic forts began much earlier. At age 20, he graduated from the Corps of Cadets in Warsaw, Poland, and after the beginning of the civil war he moved to Paris, where he enrolled in the Royal Academy of Painting and Sculpture. His goal was to become proficient in civil engineering and the strategies of Sebastien Le Prestre de Vauban, who was Europe's leading authority on forts and sieges. He returned to his native Poland after the Revolutionary War ended. After being named Commander-in-Chief of the Polish Army, he saw action in the Polish–Russian War of 1792. He returned to Philadelphia in August 1797, before retiring in Paris where he died in 1817. There are statues of him in Washington, D.C., Boston, and Detroit.

Portrait of Tadeusz Kościuszko by Karl Gottlieb Schweikart. (Courtesy National Museum in Warsaw (CC0 1.0) {{PD-US-expired}})

Fort Meigs. (Mbrickn (CC BY 4.0))

Kościuszko presciently knew that Bemis Heights, a bluff overlooking the Hudson River and adjacent to a thick tract of almost impenetrable woods, would be the perfect place for the Americans to build parapets, defensive barriers, and trenches. At West Point, the Hudson is 422 feet (128.6 meters) wide. Burgoyne arrived in September but was unable to breach the elaborate defenses. The British then attempted a flanking attack through the woods, where many were shot by Pennsylvania riflemen. Troops led by Benedict Arnold (who later tried to sell details of West Point's defenses to the British) killed and wounded 600 British soldiers. Another attack by the British farther west failed and culminated with Burgoyne's surrender after being surrounded at Saratoga.

Subsequently Kościuszko moved south, where he became chief engineer of the American army in the Carolinas. During a fight in a British fort in South Carolina, he was bayoneted in the buttocks; during the siege of Charleston, he served as a field commander.

Undeterred by setbacks at Fort Clinton, farther downstream at Montgomery, a second cast-iron chain was drawn across the Hudson River in 1778. The chain was 1,500 feet (457 meters) long. The fort's construction was formidable. The walls were 9 feet (2.7 meters) tall and 20 feet (6 meters) thick. There were three redoubts and the gun batteries that faced the south were named Forts Meigs, Wyllys, and Webb.

Construction on Fort Meigs began in April of 1778, and was completed several months later. It was a U-shaped gun battery manned by soldiers from Fort Wyllis. It was named after Return Johnathan Meigs Sr., who in 1772 was a lieutenant in a local Connecticut militia. After the battle of Lexington, he was promoted to major in the 2nd Connecticut Regiment. He accompanied Benedict Arnold on his 1,100-soldier expedition that took them through Maine to Canada. He was captured and later released.

Work on Fort Wyllis, named after Samuel Wyllis, began in April of 1778. It featured a redoubt that covered the southern approach, and a complementary gun battery consisting of three 3-pounders and two 18-pounders. The stone fort's walls were 5 feet (1.5 meters) high and 9 feet (2.7 meters) thick. There were no bombproofs. More than likely, there was a palisade and a parapet that connected the redoubt and gun battery. Although this is one of the best-preserved redoubts at West Point, unfortunately it is located within the boundaries of the Academy, thus access is restricted.

Return Johnathan Meigs Sr. is most famous for leading what is known as the Meigs Raid against the British in Sag Harbor, New York, in May 1777. Leading only 220 soldiers, he crossed Long Island Sound in 13 whaleboats to Connecticut so he could attack the British fleet at night. Twelve British ships were destroyed and 90 prisoners were taken, without American loss. General Anthony Wayne was to form the Corps of Light Infantry in July 1779 and Meigs was chosen to command the 3rd Regiment, which he led at the battle of Stony Point. The 3rd Regiment was subsequently disbanded. Meigs returned to the 6th Connecticut and then became commander of the 1st Connecticut Brigade. On January 1, 1781, after a reorganization of the Continental Army, Meigs retired as a colonel.

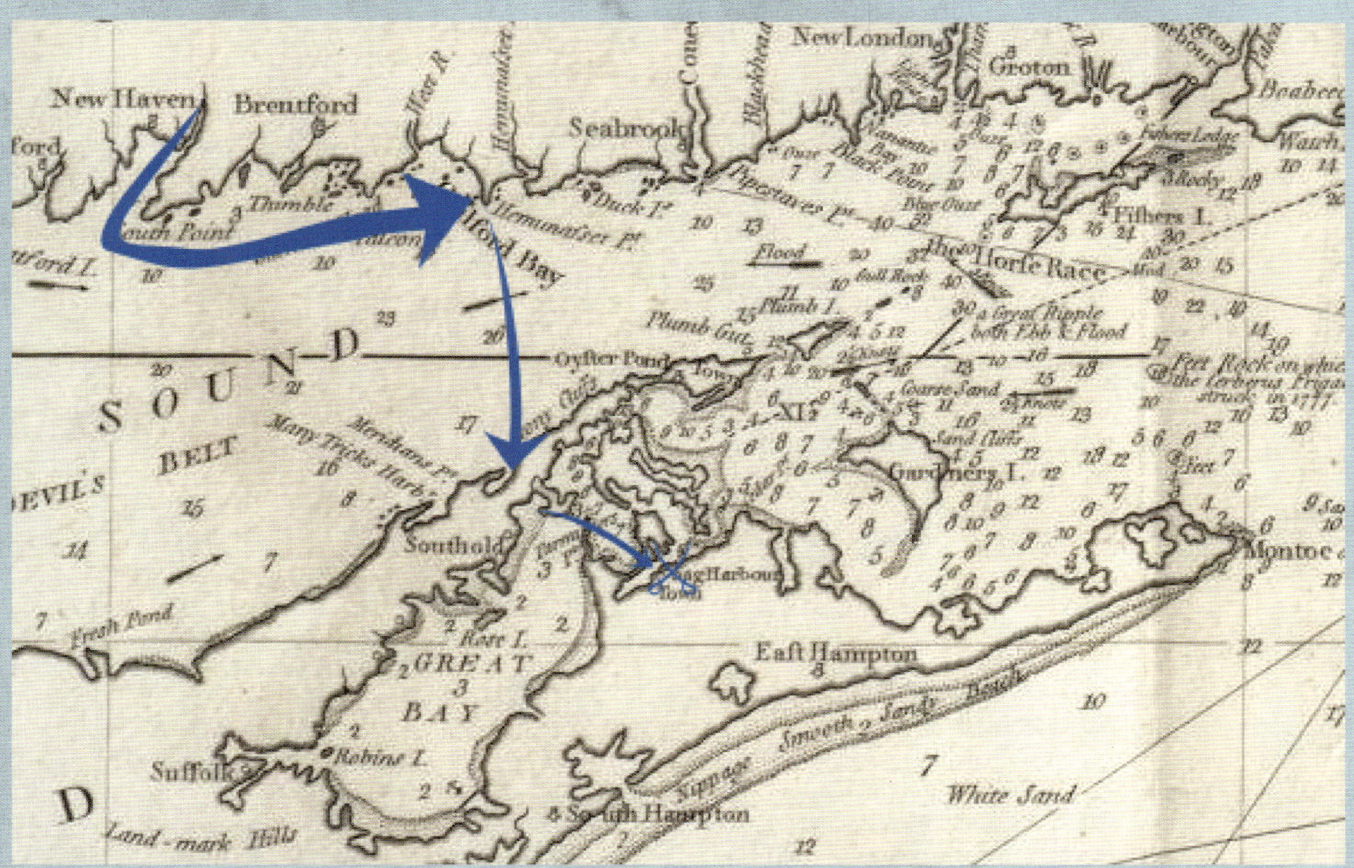

A 1794 map of eastern Long Island and the Connecticut coastline, annotated to show the route of the 1777 Meigs Raid. (Courtesy of Boston Public Library Digital Map Collection, annotations Magicpiano (CC0 1.0) {{PD-US-expired}})

Samuel Wyllis was made lieutenant-colonel in Colonel Joseph Spencer's 2nd Connecticut Regiment in 1775. Wyllis fought in the battle of Long Island and commanded the 3rd Connecticut Regiment from 1777–81, during which time he served under General Samuel Holden. The regiment remained in New York for the rest of its service. His regiment was disbanded on January 1, 1781.

Portrait of Samuel Blachley
Webb by John Trumbull.
(Courtesy www.nyhistory.org
(CC0 1.0) {{PD-USexpired}})

Samuel Blachley Webb was commanding officer of the 9th Connecticut Regiment. He was an ensign in the Wethersfield Militia Company that fought at the battle of Bunker Hill. After that, Webb was promoted to captain, brevet major, and aide-de-camp to Major Israel Putnam. In June 1776, he was promoted to lieutenant-colonel and aide-de-camp to General George Washington. Webb had the unenviable distinction of being the only aide to General Washington to be wounded three times. He rode in Washington's boat across the Delaware River on December 25/26, 1776, and shortly thereafter, on December 26, while taking orders across the battlefield during the battle of Trenton, was shot off his horse. While recovering, he was promoted to regimental colonel in January 1777. In December of that year, he was captured by a British warship. After he was paroled, he married, and returned to his regiment until it was disbanded in September 1783. When he retired, Washington promoted him to brigadier-general.

It only took two weeks and 200 soldiers to complete Fort Webb, named after Samuel Blachley Webb, in April 1778. It was an L-shaped redoubt that could hold a garrison of 375 men. It featured four gun embrasures, a bombproof, magazine, one 12-pounder, two 6-pounders, one 4-pounder, and scattered abatis. The fort was subsumed by the Military Academy. This fort no doubt secured the upper reaches of the Hudson for the Americans and did much to prevent British excursions by frigate or other vessels. The fort never fired its guns in anger.

The fort today:

Deterioration was slow, steady, and inevitable, and after the Revolutionary War, what remained of the fort was mercifully razed to allow for the expansion of the United States Military Academy. A few earthworks and some stone-based structures are all that remain, and can be seen off of Thayer Road near the soccer fields. There is no website associated with this fort. However, West Point is a charming place to visit; some parts of the Academy are open to the public, and the views of the Hudson from the bluffs as it curves south are sublime.

The British Army

The British Army command and structure was quite bewildering, in that brigade, regiment, and battalion were substitutable and that officers often led several commands. Companies were usually led by captains, but it was not unusual for them to be commanded by majors. A captain could also be in charge of a regiment that was usually led by a major.

The entire army was led by a Commander-in-Chief, and beneath him was either a major-general or a lieutenant-general. British generals never retired. In 1775, there were 119 generals in the army, but a third, approximately 39, no longer commanded in the field. Many generals were opposed to the war in America and refused to serve there. General William Howe was 111th in seniority when he took command of British forces in America.

The average size of a British regiment was 600 to 700 men, although some regiments had as many as 1,000 men, and others fell below 600 because of casualties, death, or illness. The British Army also made use of a large number of German mercenaries during the war. If a regiment

Reenactment of the land/sea battle of Hampton, Virginia. British troops look to come ashore. (watts_photos courtesy Flickr (CC BY 2.0))

was large, say over 1,000 men, it was divided into two battalions each of 10 companies and was commanded by a colonel. The two companies on the flanks were often turned into their own battalion. To make matters even more confusing, a regiment was often called a battalion even though there was a distinctive variance.

The command structure of a battalion was the same as that of a regiment and had a colonel in charge. Flank battalions got their colonel from within the regiment, while a captain in the regiment was usually a major of the battalion. Companies that were within the battalion were led by captains. Detached flank battalions, which were always made up of elite troops, had between 10 and 12 companies, and were bigger than the foot battalions that usually had eight companies. They were used almost exclusively by General Howe to lead any offensive action against the rebellious Americans. At least, that was the plan formulated by Howe while he was in Halifax, Nova Scotia. The basic organization was fashioned and rejiggered to meet the particular tactical needs in fighting an unconventional foe i.e., the Americans.

During the battle of Long Island, these flank battalions saw action. Lieutenant-General Charles Cornwallis led the 1st, 2nd, 3rd, and 4th Grenadier Battalions. Brigadier-General Alexander Leslie, whose commanding officer was General Henry Clinton, led the 1st and 2nd Light Infantry. The 3rd Grenadier and 4th Grenadier Battalions also took part in the battle. The majority of grenadier companies were authorized to have one captain, three lieutenants, five sergeants, five corporals, two drummers, two pipers and around 100 privates.

Grenadier battalions were often formed and deployed in response to a specific need. For example, the 4th Grenadier Battalion was formed in July 1776. On August 22 it was in the first wave that landed on Long Island, where it fought on August 26 and 27. It was in the first wave at Kip's Bay, landing on Manhattan on September 15, and participated in the battle of Harlem Heights. An epidemic struck the unit in October 1776 and it was disbanded. Its survivors were transferred to the 3rd Grenadier Battalion.

Grenadier, 40th Regiment of Foot, 1767. (R. H. Raymond Smythies courtesy of Historical Records of the 40th (2nd Somersetshire) Regiment)

Fort Dayton

This was an ideal location for a fort. In what is now Herkimer, located on the northern side of the Mohawk River, approximately 15 miles (24 kilometers) southeast of Utica, Palatine German immigrants settled here in 1723 and established a wood-and-earth, star-shaped fort that also contained a stone church and outbuildings located within the village of Herkimer. It was the most westerly settlement in the Mohawk Valley. The fort's construction was supervised by Colonel Elias Dayton of the 3rd New Jersey Regiment, on the same site where, during the French and Indian War, a previous fort had been built. During that war, Dayton served as a captain in the New Jersey Militia. He later became a brigadier-general. In addition to Fort Dayton, he also supervised the construction of Fort Schuyler (formerly Fort Stanwix) in what is now Rome, New York. There was also a sister fort named Fort Herkimer that was situated on the southern side of the Mohawk.

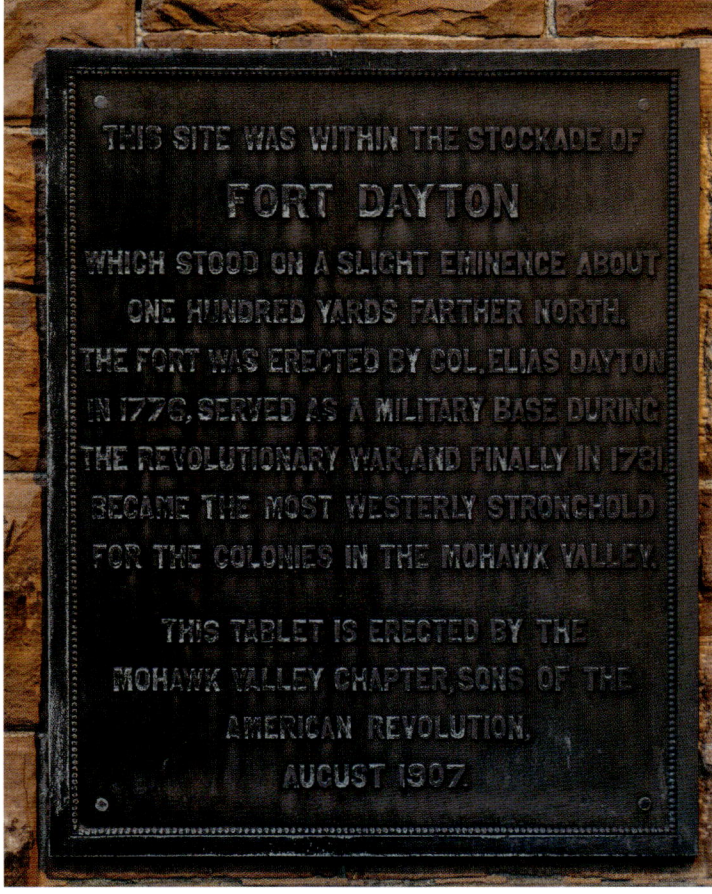

The Mohawk River is named for the Mohawk Nation of the Iroquois Confederacy. In Mohawk it was called *Tenonanatche*, or a river flowing through a mountain. It is the largest tributary of the Hudson River and has an elevation of 10 feet (3 meters), is 149 miles (240 kilometers) long, has six tributaries on its left bank, five on its right bank, and one waterfall. It is an important artery that facilitates migration and transportation through the Appalachian Mountains, between the Catskill Mountains and Allegheny Plateau to the south, and the Adirondack Mountains to the north. The area was colonized by the Dutch in 1661, who founded Schenectady about 19 miles (31 kilometers) from Albany, and later in 1712 by the British who built Fort Hunter at the convergence of the Mohawk and Schoharie Creek, approximately 22 miles (35 kilometers) upriver from Schenectady.

Plaque on the County Treasurer's Office, referencing Fort Dayton, Herkimer, New York. (Kenneth C. Zirkel (CC BY-SA 4.0))

The Mohawk River, a vital artery in the Revolutionary War. (Popular Graphic Arts courtesy of Library of Congress (CC0 1.0))

Elias Dayton was a New Jersey merchant and soldier who had served as a lieutenant during the French and Indian War and as a captain in the New Jersey Militia. Promoted to colonel, his 3rd New Jersey troops were tasked with erecting forts in New York's Mohawk Valley to defend colonists from Indian and Loyalist insurgencies. He built Fort Dayton, and also established a spy network for George Washington on Staten Island to work in conjunction with a proven American intelligence agent, John Mersereau. During Pontiac's War in 1760, he was a commander in the Detroit area. When New Jersey turned against Royal Governor William Franklin in 1775, the New Jersey's Provincial Congress named Dayton to lead the 3rd New Jersey Regiment. During that same year, he and William Alexander co-led militia forces that managed to capture a British transport off the coast of Elizabethtown. Dayton saw combat in the battles of Springfield, Monmouth, Brandywine, and Yorktown, and had the harrowing experience of having horses shot out from beneath him at Germantown, Springfield, and Crosswick's Bridge. Dayton, Ohio, is named after his son Jonathan, who was the youngest signatory of the United States Constitution.

General Elias Dayton by James Sharples. (Courtesy of Detroit Institute of Arts)

On August 4, 1777, General Nicholas Herkimer, who was fluent in German, English, and Mohawk, and approximately 900 soldiers of the Tyrone County Militia, who were garrisoned at Fort Dayton, came up to support Colonel Peter Gansevoort at Fort Stanwix, 28 miles (45 kilometers) away, that was being besieged by Barry St. Leger and his Indian, Tory (Loyalist), and British soldiers. He was ambushed in what became the battle of Oriskany.

A film loosely based on the battle of Oriskany titled *Drums Along the Mohawk* was released on November 10, 1939. It starred Henry Fonda, Claudette Colbert, and John Carridine, was produced by Darryl F. Zanuck, and was directed by the inestimably talented John Ford. The film cost $2 million and grossed over $10 million. It was filmed on location at Cedar Breaks, Utah.

The attack on German Flatts took place on September 17, 1778. Colonel Peter F. Bellinger led the 4th Regiment, Tyrone County Militia, subordinated to General Nicholas Herkimer, whose sister Delia he had married. Joseph Brant aka Thayendanegea was the Mohawk military leader who led the Loyalists and Indians in the partisan war in what was then the American frontier. Being forewarned of an impending attack by Brant's men, who numbered around 152 principally Mohawk Iroquois, the settlers who did not have the firepower to repel the attackers took refuge in the forts. Brant's men marauded through both communities and wrecked 63 homes, 63 barns, three grist mills, and one sawmill. They also drove off an untold number of sheep, horses, and cattle. The forts managed to survive but were abandoned. The church remained intact.

The fort was attacked multiple times, a testament to the value placed on it by the British and their allies. It also served as a base for troops aiding Johann Christian Schell after his stand at Schell's Bush, and for pursuing the Tory leader Walter Butler.

Schell, a veteran of the battle of Oriskany and a wealthy German, had prudently built a blockhouse whose site is not far from Schells Bush Road which interconnects with Route 169 between Little Falls and Middletown. Fort Dayton was around five miles to the south.

On the afternoon of August 6, 1781, a party of 60 Tories and Indians led by the traitor Donald McDonald, a Scotch refugee from Johnstown, attacked Schell's Bush. Such raids were classic examples of risk versus reward. Cows, steers, goats, chickens, and possibly other livestock were to be found in even modest settlements. Capturing livestock was a lot easier and more energy-efficient than hunting. A fat cow had more meat on it than a deer. Additionally, there were always rifles. Muskets could be somewhat temperamental and even fragile, so rifles were a valuable tactical asset and were in constant demand. Indians liked to plunder and scalp. A man could survive a scalping, but the Indian's preferred choice of weapon, the tomahawk, could easily lop off an arm, cause other grievous wounds, or cleave a head from a neck. Victims rarely survived. If they timed it right and met with little or no resistance, the raiders could plunder and make their escape before help arrived. However, if the inhabitants of the settlement put up a robust fight and troops from the fort arrived quicker than anticipated, the raiders had two choices: they could stand and fight a numerically superior force, or they could run away.

An average man can run 8 mph (13 kph), and if the man is running for his life, the speed increases to 12 mph (19 kph). Obviously, these figures do not take into account ethnicity, age, diet, or geographical location, nor are they an indicator of how far a man can run at the above speeds. But assuming the known distance between settlements such as Schell's Bush and the nearest fort, it would take a fit man approximately one hour to reach the fort. Once the man arrived at the fort, the alarm would immediately be sounded. It is reasonable to

General Nicholas Herkimer rallies the Tryon County Militia at the battle of Oriskany on August 6, 1777, painting *Herkimer at the Battle of Oriskany* by Frederick Coffay Yohn. (Courtesy of the Utica Public Library, New York (CC0 1.0) {{PD-US-expired}})

assume that soldiers and officers would more than likely be occupied with various chores, such as replenishing the supply of firewood, preparing meals, standing guard, and cleaning their weapons and guns. Some might even be on patrol far away from the fort. Orders would obviously be given, soldiers would have to stop doing whatever it was they were doing, put on their coats, grab their muskets, powder, ammunition, and other accoutrements such as rations and water, and fall into formation outside the fort in an orderly fashion, all of which took time.

Meanwhile, the raiders were attacking Johann Schell's sturdy blockhouse. The Dutchman defended his home successfully until dark. His wife was reloading the spare muskets, enabling her husband to continually pepper his attackers with fire. At one point McDonald tried to force the front door open with a crowbar, was wounded in the leg and, because his men were too far away to help him, Schell seized the opportunity and promptly dragged him inside. Ammunition was running low for the defenders. McDonald saved his skin by giving the defenders his own ammunition.

BATTLE OF ORISKANY, STATE OF NEW YORK.

An engraving by John Reuben Chapin of the battle of Oriskany. General Nicholas Herkimer gives orders from his saddle, on the ground after his horse was shot from beneath him. (*Ballou's pictorial*)

Seeing their leader captured, the raiders made an all-out assault on the blockhouse and thrust the ends of their muskets through the loopholes. Johann's wife promptly whacked the rifles with an ax, rendering them useless and causing the attackers to immediately retreat to a safe distance.

During the fight, Schell's 8-year-old twin sons had been captured and spirited away to Canada. Troops advanced from the fort as fast as they could, but before the Indians left, they told Schell that the fate of his sons depended on how the captured McDonald was treated. The Tory leader was taken to the fort where his wounded leg was amputated, which no doubt saved his life. The raiders had lost 11 men killed and six wounded. Schell's sons were eventually freed.

A year later, Indians who had hidden in a wheat field made another raid on Schell's Bush. Before troops from the fort could arrive, one of Schell's sons was killed and Schell himself was mortally wounded.

Long- and short-range reconnaissance patrols were a fact of life for many officers and troops who garrisoned Revolutionary War forts. Typically leaving the confines of the fort at dawn, or perhaps a little later depending on the mission they were tasked with, the men slogged for miles often in uncomfortably hot temperatures or chilling downpours. They often ventured into unknown territory, but even if they stuck to familiar trails, the threat of imminent danger was always present. In many respects a day-long trek where nothing out of the ordinary occurred could be just as nerve-wracking as getting into a fight. It was incumbent on the fort to extend its influence beyond the physical boundaries of its walls and the range of its guns, and the only way to effectively do that was to put boots on the ground and deploy them far beyond the fort.

Attacks on settlements had occurred before, notably at Andrustown in July 1778 and Little Falls and Rheimensnyder's Bush or Yellow Church in August 1781. It was an easy march from Canada to these settlements and Loyalists had no difficulty in enlisting Indian allies. These raiders knew the country well and were more than comfortable in negotiating the woods, rivers, and streams that proliferated throughout the area.

Not all patrols went as intended. Some ended badly. At the start of the war in 1776, Solomon Woodworth of Salisbury, Connecticut, enlisted as a private in the 3rd Regiment of the Tyrone County Militia and over time participated in numerous battles that saw him promoted to lieutenant in Colonel Marinus Willett's regiment on April 27, 1781. He had a well-deserved reputation for being an excellent scout, a valuable asset to the unit he was assigned to. His skill set had served him and the men he led well.

Joseph Brant aka Thayendanegea, the Mohican leader and a staunch British ally, was the scourge of the New York frontier. A painting by Gilbert Stuart. (British Museum)

45

Colonel Marinus Willett, a painting by Ralph Earl. (Bequest of George Willett Van Nest courtesy of Metropolitan Museum of Art (CC0 1.0))

On September 6, 1781, Woodworth and 46 specially chosen soldiers, augmented by six Oneida Indians, made the journey from Fort Plain (known then as Fort Rensselaer) to Fort Herkimer and from there to Fort Dayton. His unit was attached to Colonel Willett's regiment. After spending the night within the confines of Fort Dayton, Woodworth and his men left on Friday, September 7, and forded the West Canada Creek. Woodworth had been tasked with finding, engaging and hopefully destroying Loyalists and their Indian allies who were known to be in the area. It was deemed that the number of men Woodworth led was sufficient to complete their mission. They carried seven days' worth of rations, and more than enough ammunition to allow them to engage any hostile force they might encounter.

On a ridge on the eastern side of West Canada Creek, approximately 3 miles (4.8 kilometers) from Herkimer village, Woodworth and his men discovered a freshly made trail, a sure and encouraging sign that their quarry was in the area. Woodworth was an affable, easily approachable officer who did not let his rank prevent him from listening to suggestions from ordinary soldiers. It must be remembered that Woodworth had handpicked these obviously skilled and experienced men. They were not run-of-the-mill troops by any stretch of the imagination. More than one of his soldiers recommended that it would be a good idea if a runner was sent back to Fort Dayton to request Captain Garrett Putnam for additional support. Keen to engage and destroy the enemy, Woodworth declined.

A single Indian was then spotted in a deep ravine near a freshly used fire pit. Woodworth and his men charged. It was a textbook ambush. It also begs questions. Why did Woodworth and his experienced men walk into terrain that was unfavorable to them, and commit their forces before even seeing the bulk of their quarry? Was Woodworth so confident in his own abilities that prudence was overshadowed?

The West Canada Creek, at the high falls of Trenton. (Popular Graphic Arts courtesy of Library of Congress (CC0 1.0))

Lieutenant John Cement, a member of Colonel John Johnson's King's Royal Regiment, plus about 80 Cayuga, Onondaga, and Stockbridge braves led by an Onondaga chief called Daiquanda, were hidden in the copses in a semicircle around Woodworth and his men. It was a similar tactic used against General Nicholas Herkimer at the battle of Oriskany.

The end came swiftly. Woodworth and 10 of his men were slain in the first volley. The Indians charged what was left of the patrol and attacked them with tomahawks and spears, resulting in another 22 being killed, plus two of Woodworth's subordinate officers and one trooper being wounded. The survivors understandably ran for their lives. Fourteen of the original handpicked men and one wounded Oneida Indian made it safely back to the fort. Nine men were captured and taken to Canada. The attackers suffered two wounded Onondaga Indians. The next day the patrol's survivors, accompanied by Colonel Putnam's company, returned to the scene of the ambush to give their dead comrades a proper burial.

The fort today:

In 1959 a monument on Smith Road was erected by the Herkimer County Board of Supervisors. Fort Dayton and Fort Herkimer both reached the end of their lifespans and were subsequently vacated. Nothing remains of them. Fort Dayton was levelled during the construction of the Erie Canal. A memorial marker can be found at the Four Corners intersection on North Main Street in Herkimer.

The Continental Army

By way of contrast with the British Army, the American Army command structure was a much simpler, almost bespoke system. Earlier on, each of the 13 states kept a militia for local defense. In addition to the local militia were regulars called state troops, under the control of state governments. There was a third entity composed of full-time troops that formed the Continental Army and served at the national level under the authority of the Continental Congress.

In an effort to promote national unity, names were changed. For example, the 3rd Connecticut Regiment became the 20th Continental Regiment. During 1775 and 1776, regiments raised from the colonies were subsumed into the Continental Army. By December 1776, the army had approximately 120 regiments. Four were light dragoons, five were artillery, and the rest infantry.

Regiments were still defined to the borders of their state lines because it gave men an equitable path for officer promotions. Up to captain seniority was confined to his specific regiment. Field grade officers such as majors, lieutenant-colonels, and colonels claimed seniority within the state line. Congress handled general officer promotions by seniority on a national basis.

The artillery and light dragoons were run according to line rules and regulations. However, some of the older units, such as infantry regiments, were not tethered to a particular state, and were called Extra Continental Regiments. An additional 16 infantry regiments were sanctioned in December 1776, and were called Additional Continental Regiments. However, only 14 of them were actually created. In 1776, specialized units were created such as two maintenance regiments: one to guard prisoner-of-war camps, and a few mixed infantry and cavalry units called legions or partisan corps.

Regiments were the building blocks of the Continental Army. They had a solid command structure and several companies (usually eight during the first half of the war and nine during the second half, but these numbers could vary between 10 and six). A colonel normally led a

Continental Army infantry, 1779–83, by Henry Alexander Ogden. (Courtesy of Library of Congress (CC0 1.0))

regiment and was assisted by a lieutenant-colonel or a major. A captain commanded a company and his staff included lieutenants, ensigns or, for mounted units, cornets.

Regiments had an adjutant. Under him was a sergeant responsible for administrative matters, a quartermaster and a quartermaster sergeant who took care of logistics, a surgeon and his deputy (called a surgeon's mate) who attended to the wounded, a drum major and a fife major who were responsible for communications, which was a vital job prior to and during battles when voices were often drowned out by cannon or musket fire, and earlier on, a chaplain. At full strength a regiment would have eight companies and around 728 officers and regular troops, while a company would consist of 90 officers and regular troops. When engaged in combat, a regiment would be organized as a battalion and the eight companies that constituted the front line were called platoons.

As the war progressed, bigger units might engage the enemy as two battalions while companies fought as two platoons. These tactics were predicated by the limitations of the smooth-bore musket: a slow rate of fire, inherent inaccuracy, and relatively short range. When all members of the front line fired at once, they created what was essentially a giant shotgun blast. While the first line reloaded as fast as they could, the second line fired their muskets. Unlike the Europeans, who favored a three-man-deep line and bayonet charges, the American Army limited their lines to two. A brigade commanded by a brigadier-general was the result of several regiments being consolidated. In turn, more than one brigade formed a division led by a major-general.

Reenactment of Continental Army troops at Yorktown, 1781. Steven C. Berger (CC BY-SA 3.0)

Fort Decker

Fort Decker, as it is known today, in Port Jervis, New York, is represented by a stone house built in 1793 from what remained of the original Fort Decker that was constructed prior to 1760 by Frederick Haynes, a Dutch settler, as a trading post and a means of defense during the French and Indian War. Built of stone and overlapping logs, it was one and a half stories high. The fort is named after Lieutenant Marthinus Decker, the great-grandson of Jan Gerritsen Decker.

John Hathorn served on the committee that determined the location of the Hudson River chain. In 1786, September 26 to be exact, Hathorn became a brigadier-general in the Orange County Militia. On October 8, 1793, he became a major-general in the state militia. (The World War II liberty ship USS *John Hathorn* was named after him.)

A joint colonial militia force, under the command of Major Samuel Meeker of the Sussex County Militia and Lieutenant-Colonel Benjamin Tusten of the Goshen Militia, marched toward Minisink Ford to attack the

Fort Decker in Port Jervis, New York is a stone house built from the remains of the original fort by that name. The original was built before 1760, and was burned down in a raid in 1779 during the Revolutionary War. The house was built in 1793 and is the oldest building in Port Jervis, predating the town itself. (Beyond my Ken (CC BY-SA 4.0))

A facial composite portrait of John Hathorn done in 1907, 82 years after his death, by C. Brower Darst. Courtesy Albert Wisner Public Library (CC0 1.0) {{PD-US-expired}})

rampaging Brant and his Indian and Loyalist troops. The combined American force totaled 149 soldiers. Fort Decker and its garrison had suffered a crushing defeat after Brant had attacked it and left it in ruins. The attack demanded and received a rapid response.

An ambush was set up on the morning of July 22 in the hills above the Delaware River where Brant's men were crossing at Minisink Ford. Tusten was of the opinion that the marauders outnumbered the combined militia force by at least two to one. Because of that, he wanted to wait until the Americans received reinforcements from the army.

Tusten was overruled and the force marched through the night, intending to attack Brant the following day. The Americans unwisely split their force in two, hoping to outflank Brant. Before the trap could be sprung, Captain Bezaleel Tyler III of the militia prematurely fired at one of Brant's Indian scouts. Alerted, Brant then split his own force in two and quickly outflanked the Americans. Undisciplined and arguably poorly led, Hathorn's men retreated, leaving the remaining militiamen outnumbered and surrounded.

The battle lasted several hours and was eventually reduced to hand-to-hand combat. As mentioned, one of the most effective weapons the Indians used was the tomahawk that was typically made of stone and wood and consisted of a lightweight wooden handle and a thin square blade. Some had a thick spike that protruded from the back of the blade. It was a deadly weapon, an excellent tool for chopping either wood or men. Lieutenant-Colonel Tusten was killed by one such weapon during the fight. Forty-eight militiamen were killed and an additional 40 who were either wounded or captured were later killed. Most were scalped. John Hathorn managed to survive, although he was badly wounded.

Brant fared better. Only three of his men were killed, 10 wounded, and of those four did not survive long. Outraged not only by the defeat but also by Brant's atrocities, the Continental Army dispatched 3,000 soldiers to upstate New York where they burned and destroyed every Iroquois village they came upon. In late August Brant was finally defeated at the battle of Newtown.

After the battle of Minisink, Lieutenant Marthinus Decker of the Orange County Militia wrote that on July 19, 1779, a day long to be remembered by the people living in the valley of the Neversink, the Indians again made their appearance on the frontier and burned a number of houses and barns and destroyed much

The monument at the site of the battle of Minisink. (Charles Fulton courtesy of Flickr (CC BY-SA 2.0))

property. The alarm was given and the Goshen Regiment was called out. Captain Kortright's company pursued the Indians as far as Beaver Brook, where they overtook them and engaged. The militia were defeated. Colonel Tusten was killed with a fatal tomahawk chop to his head along with a number of other officers and men. Captain Wood was taken prisoner. Samuel Meeker was wounded and retreated to the relative safety of his house.

The fort today: The Fort Decker Museum of History is located at 127 West Main Street, Port Jervis, NY 12771. There is also a stone house built from what remained of Fort Decker; that structure is now a museum. It is listed on the National Register of Historic Places. The previous iteration had been reduced to rubble and rendered virtually uninhabitable. The current house is obviously not a literal recreation or representation of the original fort but is more of a memorial. In Sullivan County is the Minisink Battleground County Park that has a visitor center, nature trails, picnic areas, and monuments. The park is located on Sullivan County Route 168 about a mile northeast of New York State Route 97 at Minisink Ford.

www.scenicworlddelawareriver.com/fort-decker-port-jervice gives information not only about the fort but about how it is connected with the local community.

Flagstaff Fort

This fort was constructed in June 1776 on Signal Hill at the strategically important Narrows on Staten Island, on the site of an earlier blockhouse built in 1663 that was itself preceded by a similar structure erected by the Dutch settler David Pieterszen de Vries in 1636. The Dutchman's fort was wrecked in the Peach Tree War of 1655.

The fort was rebuilt by the British in June 1776 and, three years later in July 1779, a redoubt with gun platforms that could accommodate 26 cannons was built. The British wasted no time in adding to the impressive array of firepower at the fort. Two months later, six huge 24-pounders and 4 complementary 18-pounders joined the guns already in place. Three years later, in 1782, the fort had five bastions and several barbette batteries. In 1783 the British departed at the end of the Revolutionary War.

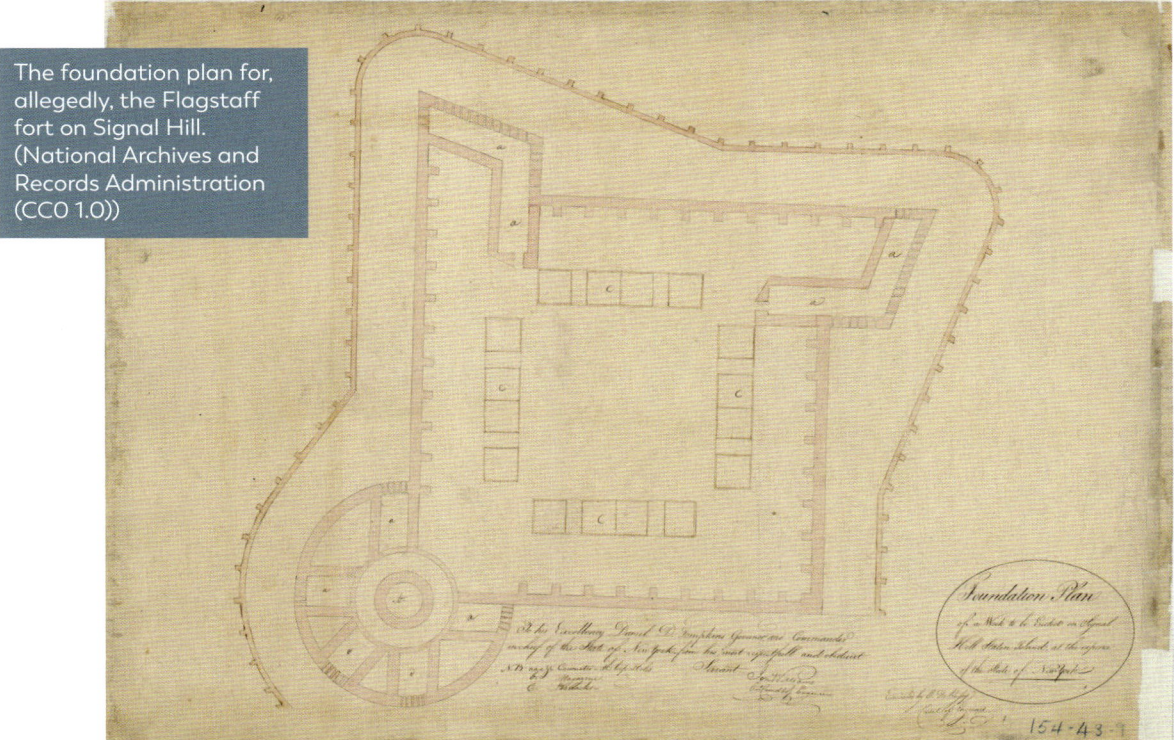

The foundation plan for, allegedly, the Flagstaff fort on Signal Hill. (National Archives and Records Administration (CC0 1.0))

The Flagstaff fort was demolished in 1806 when Fort Tompkins was built, reportedly enclosing the original blockhouse with limestone. Tompkins was later amalgamated into Fort Wadsworth that is now part of the Gateway National Recreation Area of the National Park Service.

The fort today:

For more information, see the chapter on Fort Wadsworth.

The navies

The command structure of the British navy was relatively straightforward and well defined. Commissioned officers, who unlike their army counterparts did not purchase their commissions, served under the captain. Most started their career as midshipmen in their teenage years and learned the fundamentals of seamanship, navigation, and leadership. Next came a lieutenant's exam and if the applicant passed, he was awarded a commission. Becoming a captain was trickier and often involved political or social connections or a display of courage in battle.

The next ranks down were warrant and petty officers who were responsible for the actual running of the vessel. Included in these ranks were the ship's gunner, surgeon, carpenter, and the ship's master. Every ship carried a contingent of marines used for amphibious landings.

While the vast majority of seamen who sailed and fought on British vessels were volunteers, the need for men increased exponentially with the increase in warfare and impressment was frequently used to complete a ship's crew. Bearing a royal warrant, a press gang led by an officer would comb through a port city, authorized to take or impress any man with even rudimentary

A Royal Navy press gang in action, an illustration by Charles Joseph Staniland.

An early American privateer around the turn of the century. (Courtesy of Museum of Fine Arts, Boston)

sailing familiarity between the ages of 15 and 55. Naval seamen would often board merchant ships at sea and literally take crewmen if they were shorthanded for whatever the reason. Pay on a naval vessel was a lot less than a man could earn on a merchant ship, so naturally this was very unpopular. In some port cities impressment led to riots and it was one of these protests that contributed to the American Declaration of Independence.

Although the fledgling American navy posed no real threat to their British counterparts, they instead concentrated on British shipping and authorized over 1,000 privateers to attack British merchants. During the Revolution over 2,000 merchants were captured, which did much to turn British opinion against the war.

Fort au Fer

Rouses Point—once Rouse's Point, the apostrophe has long gone—is a small wooded peninsula on Lake Champlain. Approximately a mile south of Rouses Point is where in 1775 the British built this blockhouse-style fort. It was of average size, measuring 40 feet (12 meters) by 50 feet (15 meters). Originally constructed of stone, brick barracks and a stockade were later added. It was fortified with cannons and entrenchments. Fish and game were plentiful. Officers most likely hunted and fished and as a result ate relatively well. Even during the summer months there was a breeze off the lake that kept temperatures and tempers tolerable.

General John Burgoyne and his staff occupied the fort during the general's Saratoga campaign in 1777 and it remained in British hands until 1796, following the 1794 Jay Treaty that took effect on June 1. This treaty ended, for all intents and purposes, British occupation of Lake Champlain.

The treaty was an agreement between the United States and Great Britain that recognized a foundation that would allow the United States to develop a robust national economy, thus assuring its commercial success. Britain also agreed to quit the Northwest Territory no later than June 1, 1796, to pay the United States for its despoliations of American shipping, and not to show prejudice against American commerce. The treaty additionally acknowledged that the Mississippi River was available to both countries, that ports in the United States would not outfit privateers hostile to Britain, paid Americans who had sustained debts to British importers before the Revolution, and created commissions that would over time determine boundaries between the United States and British North America in the Northwest and Northeast. The treaty's ratification formed the basis for much of the present day's principle of arbitration.

A map of Rouse's Point and environs, Lake Champlain. (National Archives and Records Administration (CC0 1.0))

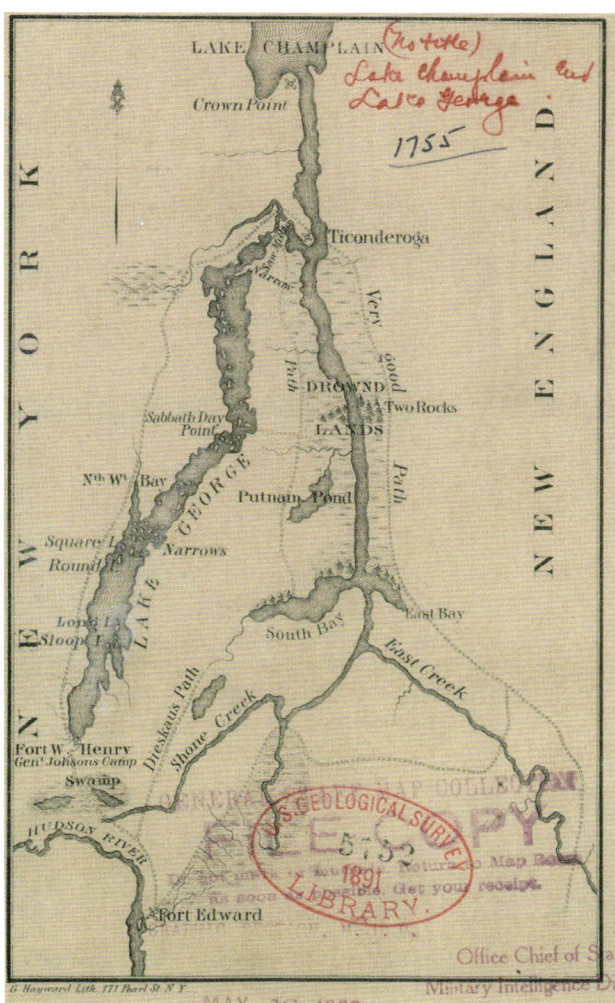

A map of Lake George and the southern end of Lake Champlain, with Crown Point visible. (National Archives and Records Administration (CC0 1.0))

In 1774, the British built a fort known as the White House (in May 1776, a small American force took control of the fort). The garrison was no doubt keeping as close an eye as possible on events happening not that far to the north. On December 31, 1775, with American forces attempting to invade Canada, the battle of Quebec began. The British forces were led by Major-General Guy Carleton and Brigadier-General Allan Maclean.

The Americans were led by Major-General Richard Montgomery, who had previously served in the British Army during the French and Indian War. In June 1775, he was given the rank of brigadier-general and assumed leadership of the invasion of Canada when the acting commander Phillip Schuyler fell ill. In relatively short order, Montgomery captured Port St. Johns; another triumph followed in November 1775 when he conquered Montréal before advancing on Quebec City, where his forces joined up with those of Benedict Arnold.

By all accounts, the battle of Quebec was an unmitigated and complete disaster for the Americans and was the first major defeat suffered by the Americans in the Revolutionary War. Montgomery, Arnold, Daniel Morgan, and James Livingston commanded a force of 1,200 men against Guy Carleton and Allan Maclean, representing the British, who fielded a force of 1,800 troops.

Montgomery and Arnold split their combined force in two so they could converge on the lower part of the city before scaling the upper city's defensive walls in a two-pronged attack. Early in the battle Montgomery was unfortunately killed by cannon fire, causing his men to lose heart and turn back; this effectively ended the attack that initially had shown much promise. However, unlike their compatriots, Arnold's men persevered and after bitter fighting managed to penetrate the lower city. Again, misfortune struck. Arnold was injured and incapacitated and Morgan, who had assumed command in his absence, became hopelessly ensnared in the lower city and was forced to surrender. In spite of these setbacks, a feeble blockade was initiated and managed to continue until the spring when British reinforcements arrived and ended the blockade. In the battle, the British lost 19 men killed and wounded, while the American losses were 84 men killed and wounded, and 431 captured.

Robert Rogers was born in 1731 in Methuen, Essex County, Massachusetts. He founded Rogers' Rangers to contest the area around Lake Champlain and Lake George in the French and Indian War (1754–63). He was to defeat the French as Scales Point, forcing them back into Canada. His unconventional tactics, operating behind enemy lines, saw him destroy several French military posts along the Richelieu River in Quebec. His nickname was "The white devil." Rogers' Rangers saw action in the battle of Snowshoes, the battle of Carillon, St. Francis Raid, Thérèse Raid, the Montréal campaign, Pontiac's War, the battle of Bloody Run, and the battle of Mamaroneck.

Equivalent to today's mobile strike force—the U.S. Army Rangers is direct descendant of Rogers' Rangers—Rogers used innovative techniques such as sleds, rudimentary snowshoes, and even ice skates to cross frozen rivers. Tactics like these allowed him to raid French towns and military installations when the occupants least expected it, hunkered down as they were where temperatures routinely plunged well below zero.

Between 1755 and 1758, Roger's Rangers increased in size to 12 companies plus sizable groups of Indians who were sympathetic to and allied with the British. Rogers was understandably kept apart structurally from regular British troops and was allowed to function with a large degree of autonomy. In 1759, Rogers was tasked by Major-General Jeffrey Amherst, the new Commander-in-Chief of the North American British forces, with an expedition against the Abenakis at Saint Francis in Quebec. Rogers continued to fight in the northeast and even the Great Lakes for years. His final act as commander was his occupation of Fort Michilimackinac and Fort Joseph during the spring of 1761.

Major Robert Rogers. (Thomas Hart (publisher) Johann Martin Will (artist) courtesy of Anne S. K. Brown Military Collection (CC0 1.0) {{PD-US-expired}})

"To range the woods" was the maxim of Rogers' Rangers. (Army Artist Team XXII courtesy U.S. Army Center of Military History (CC0 1.0))

The Death of General Montgomery in the Attack on Quebec, December 31, 1775 by John Trumbull. (Courtesy Yale University Art Gallery (CC0 1.0) {{PD-US-expired}})

Arnold's column is broken by Canadian militia and British troops in fierce street fighting during the battle of Quebec, a painting by Charles William Jefferys. (William Wood (CC BY-SA 3.0) {{PD-US-expired}})

Sir Guy Carleton, the victor of Quebec by Baron H. de Dirckinck Holmfeldt. (Courtesy Jean Gagnon (CC BY-SA 3.0) {{PD-US-expired}})

What remains of Fort St. Frédéric, Crown Point. (Petersent)

James Livingston was oddly enough a colonel in the 1st Canadian Regiment, living in Chambly, Quebec, and working as a grain merchant when the Revolutionary War began. During the invasion of Canada, he assumed command of a regiment. He fought at the siege of Fort St. Jean, the battle of Quebec, the battle of Trois-Rivières, the battle of Saratoga, and the battle of Rhode Island. It was his involvement due to luck, happenstance, or simply being in the right place at the right time for which he is remembered.

John André was born in London and fluent in English, Italian, German, and French. By all accounts he was a suave raconteur who could not only draw, paint, and make silhouettes but who also sang and wrote verse. At the age of 20, in 1774, he was a lieutenant in the 7th of Foot (Royal Fusiliers). In November of 1775 he was captured by General Richard Montgomery at Fort Saint-Jean and held prisoner in Lancaster, Pennsylvania. In 1776 he was part of a prisoner exchange and was promoted captain in the 26th Foot (Royal Fusiliers) and two years later to major. In 1779 he became head of the British Secret Service in North America.

In 1780 he began negotiations with Benedict Arnold's wife Peggy Shippen, the go-between for her husband who had agreed to surrender his command at West Point to the British. Livingston was in command of Verplanck's Point on the Hudson River in September 1780 when his troops fired on the British sloop-of-war *Vulture*. It just so happened that on board the *Vulture* were both Arnold and André. Mindful of the importance of its two passengers, the sloop prudently retreated downriver pursued by an American longboat that eventually gave up the chase. Once the Americans had been given the slip, Arnold gained the chance to have a small, prearranged boat meet the sloop; this took John André ashore where he changed into civilian clothes, with a Hessian overcoat and a fake passport that identified him as John Anderson. Everything needed to facilitate André's safe deposit on shore and the passport had been furnished by Arnold.

André now found himself marooned on shore with a horse—it is unclear where this came from—wearing for the most part civilian clothes and travelling with a fake passport that hopefully would pass casual inspection. The combination of the two were to prove to be fatal errors of judgement. Had he remained in his regular army issue uniform when subsequently arrested, he would have been considered a prisoner of war and treated as such. Wearing civilian clothes automatically labeled him a spy. There would be little even the glib André could say or do to mitigate the dire circumstances he found himself under.

At 9 a.m. on September 23, three militiamen stopped André and, after a brief interrogation during which they realized his passport was fake, arrested him. He was taken to the headquarters of the American army at Tappan where the post commander, Lieutenant-Colonel John Jameson, wanted to send him back to the custody of Arnold. André was a whisker away from being free. Again, fate conspired against André when the head of the Continental Army Intelligence, Major Benjamin Tallmadge, intervened. This intervention was motivated by intelligence that strongly indicated that a high-ranking American officer was planning to

John André. (Courtesy Tower of London (CC0 1.0) {{PD-US-expired}})

Benedict Arnold, a painting by Thomas Brown. (Courtesy of Anne S. K. Brown Military Collection (CC0 1.0) {{PD-US-expired}})

go over to the British. However, Tallmadge didn't know exactly who it was, and because he did not want to be perceived as thinking his commanding general (Arnold) was a traitor, which would have instantly ruined his career, Jameson sent Arnold a letter that informed the latter of the situation. Arnold got the letter while eating breakfast, excused himself from the table, and disappeared.

André was taken to Wright's Mill in North Castle before being transferred across the Hudson to the American army headquarters at Tappan. Imprisoned in a tavern known as the '76 House, he confessed to his true identity. Acting on the direct orders of George Washington, senior officers investigated the matter. On the board were Major-Generals Nathanael Greene, Lord Sterling, Arthur St. Clair, Lafayette, Robert Howe, and Friedrich von Steuben, and Brigadier-Generals Samuel H. Parsons, James Clinton, Henry Knox, John Glover, John Paterson, Edward Hand, Jedediah Huntington, and John Stark. Judge Advocate General John Laurance presided over the proceedings. André claimed he had not planned to be behind the American lines, that as a prisoner of war it was his right to escape wearing civilian clothes, and that he was suborning an enemy officer. To his credit, he did not pass blame onto Arnold. André was given a fair trial, found guilty of spying, and subsequently hanged, while Arnold was not long in following him to the gallows.

In the 1730s the French built a massive fort at Crown Point on Lake Champlain that years later became the border between New York and Vermont. The fort was named Fort St. Frédéric. Its limestone walls were 12 feet (3.7 meters) thick; it had four casements filled with cannons and was garrisoned primarily by soldiers from the Compagnies franches de la marine (marines). After it had been assaulted twice, the French demolished it in the summer of 1759.

In that same year, British troops and men from the New England colonies built at the same location what was then the largest earthen fort in the United States and named it Fort Crown Point. Sir Jeffery Amherst commanded the British Army at the time, but Israel Putnam was the man who supervised the construction.

When the French and Indian War, ended the British left a small garrison that readily succumbed, on May 12, 1775, to American forces led by Captain Seth Warner and 100 members of a Patriot militia called the Green Mountain Boys. The Americans captured 111 valuable cannons and sent 29 of them to Boston where they were desperately needed for the defense of Boston Harbor. Benedict Arnold used the fort as a staging ground for

A north view of Fort St. Frédéric/Fort Crown Point. (Courtesy of New York Public Library (CC0 1.0))

his navy on Lake Champlain. The British took control of the fort in 1777after the failed American invasion of Canada, but deserted it in 1780 when it was left to slowly deteriorate.

Valcour Bay is a slender channel on Lake Champlain; it lies between Valcour Island and New York. Valcour Island is near Plattsburgh, New York, not far from the Canadian border. It was there on October 11, 1776, that a naval battle known as the battle of Valcour Island, or the battle of Valcour Bay, took place. It was one of the first naval engagements fought by the fledgling United States Navy. The antagonists were Guy Carleton and Thomas Pringle for the British, and Benedict Arnold for the Americans.

The experienced British force consisted of one sloop, two schooners, one radeau, one gundalow, and 28 gunboats, while the Americans committed four galleys, two schooners, one sloop, and eight gundalows to the battle. The Americans were outnumbered by slightly more than two to one.

A gundalow was a sailing barge not unlike a scow that uses a single lateen sail and tidal currents for propulsion. They could be up to 70 feet (21 meters) long and instead of a fixed keel, had a pivoting leeboard. During the battle they carried cannons of various sizes. They were obviously not the most stable of platforms on which to fire even a small-bore cannon that still had a considerable recoil. A radeau was nothing more than an armed scow that could be rigged in a number of different ways and was basically a floating gun battery. They were slow and not particularly maneuverable.

What remains of Fort Crown Point. (Poster (CC BY-SA 2.5))

British Admiral John Schank was on board the 180-ton *Inflexible* that carried 18 x 12-pounders. The vessel had been taken apart in Quebec City and moved upriver in sections. It took part in the latter stages of the battle. Captain George Scott was in charge of the *Thunderer* that wasn't involved in the fight but was heavily armed with six 24-pounders, six 12-pounders, and two howitzers. John Starke captained the schooner *Maria* that was the flagship for Guy Carleton and Thomas Pringle. Although it carried 14 guns, it chose not to participate in the battle. The schooner *Carleton*, captained by James Richard Dacres, was armed with 12 guns, while the gundalow *Loyal Convert* had seven guns and was commanded by Edward Longcroft. The single-masted gunboats had two small-bore cannons each.

Benedict Arnold was on his flagship *Congress*, a row galley captained by James Arnold. James Smith commanded the hospital ship *Enterprise*, a sloop armed with 12 guns. Also carrying 12 guns was the schooner *Royal Savage* (Captain David Hawley). The row galley *Washington* had 10 cannons and was captained by John Thatcher. Another row galley, the *Trumbull*, also had 10 guns and was captained by Seth Warner. A third row galley, the *Lee*, had six guns and was captained by John Daviss. The American gundalows were masquerading as gunboats in an attempt to fool the British.

The three militiamen who arrested John André—John Paulding, David Williams, and Isaac Van Wert—were later each given a pension of $200 a year. They had counties in Ohio named after them and their memories are immortalized by a monument positioned where they apprehended André. It is located in Patriot's Park on U.S. Route 9, on the boundary between Tarrytown and Sleepy Hollow in Westchester County, New York. In 1982 it was added to the National Register of Historic Places.

Arnold knew he was outmanned and outgunned. Before the war he had sailed ships both to the West Indies and Europe and so seamanship wasn't an issue. He therefore carefully chose where he wanted to engage the British fleet. After transferring his flag to the row galley *Congress*, he chose a rock-strewn, slender body of water that was between the western shore of Lake Champlain and Valcour Island. Arnold hoped that the larger

British fleet would have trouble navigating in such a confined space and would therefore be unable to bring its guns to bear on him in a proper fashion. Also, many of his captains were relatively inexperienced and thus could not be counted on to execute complicated naval maneuvers while possibly under cannon fire. Even the most experienced captains were under enormous stress during a battle. The proper positioning of their ships in relationship not only to the enemy's but also to their own fleet depended on maneuverability, wind, and execution of orders that were often shouted over cannon and musket fire. Smoke often obscured views and on ships under fire there was organized chaos. Wounded men had to be tended to and guns had to be reloaded as quickly as possible, all made more difficult by the movement of the ship.

The British fleet, augmented by 50 unarmed support ships, entered Lake Champlain on October 9 and guardedly made their way south looking for the American fleet. A day later, on October 10, the British dropped anchor that night, approximately 15 miles (24 kilometers) to the north of Arnold's ships. The British still didn't know where the Americans were.

The following day, taking advantage of favorable winds, the British cautiously sailed south. Arnold, who knew what he was doing, having established a crescent-shaped firing line dispatched the *Congress* and the *Royal Savage* in a sortie, whose aim was to get the attention of the British fleet and with luck draw them into a position that was unfavorable to them. Unfortunately, due to strong headwinds, *Royal Savage* had to be beached on the southern tip of Valcour Island.

The fight began in earnest. The American line was first reached by hard-charging British gunboats, which opened fire as soon as they were within range of the *Carleton*. Unfavorable winds stalled *Thunderer* and *Maria*

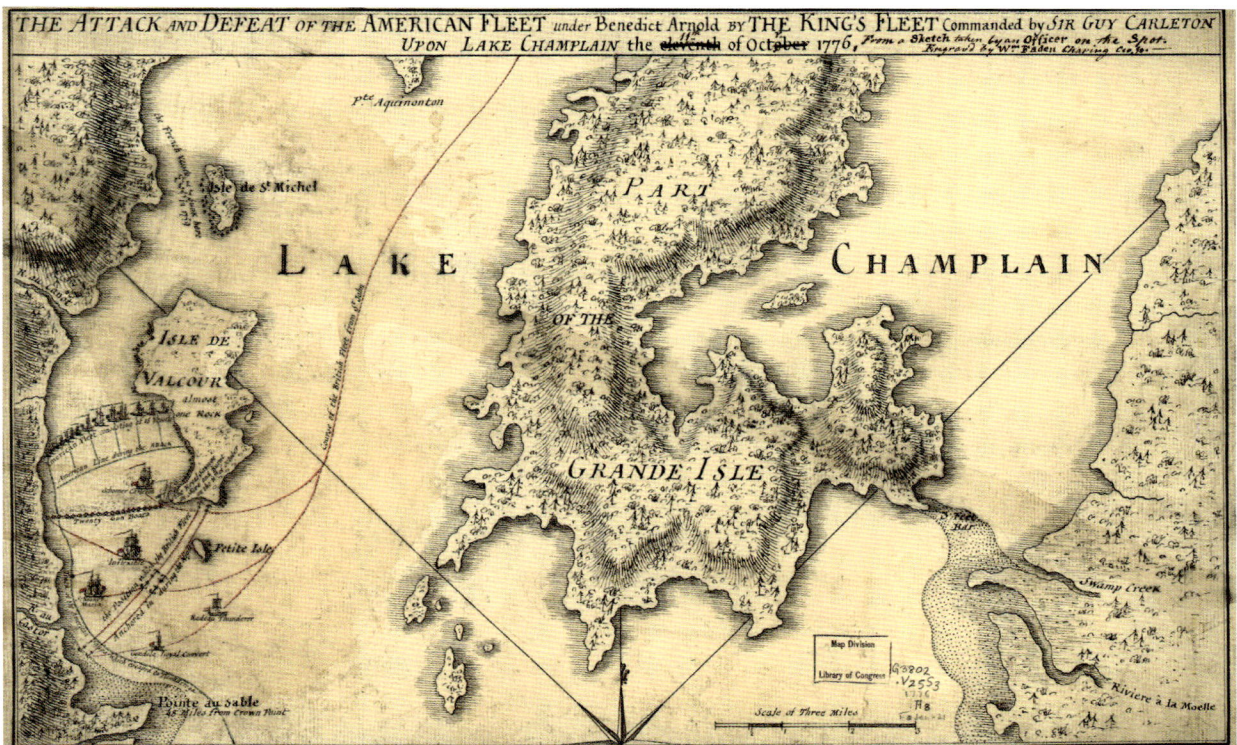

The battle of Valcour Island. (Courtesy of Library of Congress (CC0 1.0))

A view of the battle of Valcour Island, fought on October 11, 1776, by the artist Henry Gilder. Annotated ships: *Carleton, Royal Savage* (1775), *Inflexible, Maria, Royal Convert, Thunderer* radeau. Key: (a.) Cumberland Head; (b.) Cumberland Bay; (c.) I. de Valcour; (d.) Petite Isle; (e.) Grande Isle; (f.) Rebel Fleet; (g.) *Carleton* schooner; (h.) *Royal Savage* aground; (i.) Line of gun boats; (k.) *Inflexible*; (l.) *Maria*; (m.) *Royal Convert*; (n.) *Thunderer* radeau. (Courtesy Royal Collection)

in their tracks. *Inflexible* eagerly got into the fight. Shortly after noon, at around 12:30 p.m., both sides were fully engaged, firing broadsides and cannonades at each other repeatedly for the rest of the afternoon. The noise was deafening. Officers screamed orders. Gun crews reloaded as fast as they could while they adjusted the firing angles on their guns. Sometimes ships on both sides were obscured by smoke. Guy lines were taut. Seamen scrambled on decks as their captains tried to maneuver their vessels into more advantageous positions. The ships on both sides were strained to their maximum capabilities. The larger vessels had a wide turning radius and even under full sail their top speed was less than 10 knots per hour.

Revenge was hit multiple times as was *Philadelphia*, which was so badly damaged that she eventually sank around 6:30 p.m. *Carleton* received a shot that snapped the line that held her broadside in the correct position. As a result, she was critically damaged and took a lot of punishment before she could be towed out of range of the American guns. Midshipman Edward Pellew took command after the ship's officers and captain were seriously wounded. A British gunboat exploded when a lucky shot hit her magazine and mangled bodies flew into the air.

Toward dusk, the *Inflexible* finally cruised into action and began firing salvo after salvo with her big guns. Her 12-pounders had a maximum range of 1,969 yards (1,800 meters). The effective range for solid shot was 984 yards (900 meters) and 656 yards (600 meters) for canister, and a well-trained crew could fire one round per minute. The continual barrage was enough to cause Arnold to order what was left of his fleet to retire. To make matters even worse for the Americans, the British had infiltrated Indians onto the shore of Lake Champlain and also on Valcour Island to prevent the Americans from seeking safety on land. Arnold was running out of options.

In a desperate, last-ditch attempt to save what was left of his fleet, Arnold made a frantic dash to Fort Crown Point, 35 miles (56 kilometers) south of his position. During the night, he and what remained of his fleet somehow managed to sneak through a one-mile gap between the British fleet and the Indians on the western shore of the lake. Morning saw the Americans 8 miles (13 kilometers) to the south.

Valcour Bay today. (Daniel Case (CC BY-SA 3.0))

The American fleet was in bad shape. The *Providence* and *Jersey* had been sunk or burned at Schuyler Island. The western shore was where the *Lee* was beached. By the following morning, after fighting contrary winds and leaky vessels, the Americans were more than 20 miles (32 kilometers) from Crown Point but the British fleet could ominously be seen on the horizon and soon attacked again. The *Congress* and *Washington* were hit, the latter to the extent that it was unable to continue. Smaller ships took shelter and hid in a small bay on the Vermont side of the lake, where they were put to the torch as was Arnold's flagship, *Congress*. The crews of these vessels slogged overland to Crown Point where they met up with the *Trumbull*, *New York*, *Enterprise*, and *Revenge*. The *Liberty* was newly arrived with much needed provisions from Fort Ticonderoga.

Arnold's original plan was not to necessarily win a naval battle against the British, which he knew to be unrealistic, but rather to delay or even prevent the British from reaching the upper Hudson River valley. In that regard he succeeded. He quit Crown Point, which was occupied a day later by the British whose reconnaissance parties got as far as 3 miles (5 kilometers) from Fort Ticonderoga. When the first snows began to fall, Carleton prudently withdrew to his winter quarters.

The losses suffered in the battle were in almost direct proportion to the number of ships each side committed to the fight. The British lost 40 men either killed or wounded, plus one gunboat destroyed and two gunboats sunk. By way of contrast, the Americans lost 80 men killed or wounded and 120 captured. One schooner was destroyed, one galley was destroyed, two galleys were captured, three gundalows were destroyed, three gundalows were sunk, and one gundalow was captured. Britain ruled the waters of Lake Champlain and in 1774 erected a fort they named the "White House." After the war, this stronghold was occupied by the British for 20 continuous years. From here, the lake was patrolled by Captain John Steele and his ever-reliable gunboat *Maria*.

After the battle, the Americans lost control of Point au Fer. In July 1777, half of General John Burgoyne's army camped there prior to the Saratoga campaign. Fort au Fer was razed by fire in 1808 by French refugees. During the War of 1812 the site was used as a lookout post by the U.S. Army.

The fort today:

A historic marker can be found at 534 Point au Fer Road and another one is at 548 Point au Fer. The site of the naval battle is now a National Historic Landmark. The *Philadelphia*, raised in 1935, is also a National Historic Landmark. The underwater site of the *Spitfire* was located in 1997 and is on the National Register of Historic Places.

Fort Defiance

Located on an island in New York Bay, Fort Defiance was the westernmost stronghold along Brooklyn Heights. It was tasked with defending Upper New York Bay from possible infiltrations by the Royal Navy. The fort featured five decent-sized cannons, a defensive wall, and several earthworks. It saw action during the battle of Long Island in late August 1776. Israel Putnam, commanding 1,000 men, constructed the fort in a marathon, nonstop effort during one night in April, an unheard-of feat. He built three redoubts on the small island, connected them with trenches, and defended the island's southern side with an earthwork.

Putnam's little fort was put to the test on July 12, 1776, when Admiral Howe sent two ships, HMS *Phoenix* and HMS *Rose*, to run the gauntlet toward New York City. The *Phoenix* was a fifth-rate frigate that carried 44 guns and was captained by Hyde Parker. On her lower deck she had 20 x 18-pounders, her upper deck was armed with 20 x 9-pounders, and her quarterdeck had four 6-pounders. Crewed by 280 officers and ordinary seamen, the ship was essentially a floating fort. The *Phoenix* had been busy. In late June 1776, she was at anchor off Sandy Hook, New Jersey, where she seized three American vessels and disrupted an American assault on a lighthouse adjacent to Sandy Hook. On July 12 *Phoenix*, along with the *Rose* and three smaller vessels, bombarded New York for two uninterrupted hours. The *Phoenix* then harried American positions along the Hudson River, again acting with relative impunity before she withdrew to anchorage off Staten Island on August 16. The ship's demise took place on the night of October 4, 1780, when she sank during a hurricane that disabled Britain's entire fleet in the West Indies. The *Rose* was smaller and not as heavily armed. She was a sixth-rate post ship that carried 20 x 9-pounders and was crewed by 160 officers and seamen. Commanded by Sir James Wallace and acting with

HMS *Phoenix* and HMS *Rose* engaged by American fire ships and galleys, August 16, 1776, engraved from the original picture by D. Serres from a sketch by Sir James Wallace. (Courtesy of New York Public Library (CC0 1.0)

Thomas Mitchell's dramatic painting of the British fleet forcing a passage of the Hudson River, October 9, 1776. It shows HMS *Phoenix*, *Roebuck*, and *Tartar*, accompanied by two smaller vessels, forcing their way through a cheval-de-frise on the Hudson River with the Forts Washington and Lee and several batteries on both sides. (Royal Museums Greenwich)

virtual immunity in concert with the *Phoenix*, she brazenly ventured far up the Hudson River, bombarding Fort Defiance and other installations as she went. Her activities were contributory in forcing General George Washington and his troops out of New York City. Wallace was subsequently knighted for his involvement. The *Rose* was scuttled on September 19 in Savannah, Georgia.

Fort Defiance's cannons fired on HMS *Phoenix* and HMS *Rose* but had little effect. It was akin to throwing a tennis ball at an elephant. General Putnam's improvised and hastily constructed fort did its best, but the more heavily armed British ships sailed impudently and almost imperiously past the fort. Cannons on Governors Island and Fort George also took shots at the two British ships, again with no effect. New York City was then bombarded by the opportunistic HMS *Phoenix* and HMS *Rose*. They proceeded to sail unimpeded up the Hudson River so they could successfully blockade the crossing at Tarrytown.

Another ship involved in this fight, and one that came under fire from Fort Defiance, was HMS *Roebuck*, a model of the new two-deck, fifth-rated ships. Armed with 20 x 18-pounders on her lower deck, 22 x 9-pounders on her upper deck (that were later upgraded to 12-pounders), and two 6-pounders on her quarterdeck, she was crewed by between 280 and 300 officers and men. She was designed by the famous naval architect Sir Thomas Slade as an improvement on the *Phoenix* class. Fort Defiance had better luck with this vessel when it stalled in Buttermilk Channel: the American guns managed to damage her to the extent that she had to leave the fight and retire to anchorage. However, in return, Fort Defiance's redoubts had been heavily damaged and, for all intents and purposes, the fort was no longer a viable installation.

Like other forts, Fort Defiance was built in record time as a reaction to an immediate threat. Its purpose was to thwart, deter, or negate the threat. When the threat no longer existed, neither did the fort. After the war she was abandoned and her embankments leveled.

The fort today:

Nothing remains of the fort. Ironically, she disappeared almost as fast as she was created.

A cannon left behind by the British at Fort Franklin at the end of the Revolutionary War. (DeBevoise, C. Manley courtesy of New York Public Library (CC0 1.0))

| Fort Franklin

Built by the British in 1778 along the north shore of Long Island, this fort and others were tasked with implementing British occupation, initiating assaults across Long Island Sound, and guarding Manhattan from attacks from the east. Strategically placed on a bluff on Lloyd's Neck which jutted northward from Huntington, Fort Franklin was the jewel in the crown of these forts. It was named after Benjamin Franklin's Loyalist son, Sir William Franklin.

The square earthwork fort surrounded by abatis also controlled entrance from Long Island Sound to Oyster Bay and Cold Spring Harbor. British plunderers using whaleboats regularly terrorized American towns across Long Island Sound in Connecticut and were protected by Fort Franklin's guns. The garrison was commanded by the 3rd Battalion of Major-General Oliver De Lancey's Tory regiment, the New York Provincial Militia. Three years later, the Associated Loyalists made it their base of operations. Other detachments such as the Loyal New England Regiment and militias from Rhode Island filtered in and out of the fort. The garrison strength fluctuated between 500 and 800 soldiers.

The Americans wanted desperately to seize the fort and an attack was launched by 27-year-old Major Benjamin Tallmadge of the 2nd Connecticut Dragoons, whose objective was to expel the British from Suffolk County where they had access to ample supplies of fresh meat, grain, and forage. On September 5, 1779, at around 10 p.m., Tallmadge and a force of around 130 refugees, boatmen, and dragoons seized outbuildings consisting of houses and huts. The small encampment was seized without incident. The raiders next tried to seize huts used by enemy whaleboat men. Loyalists responded with volleys of musket fire.

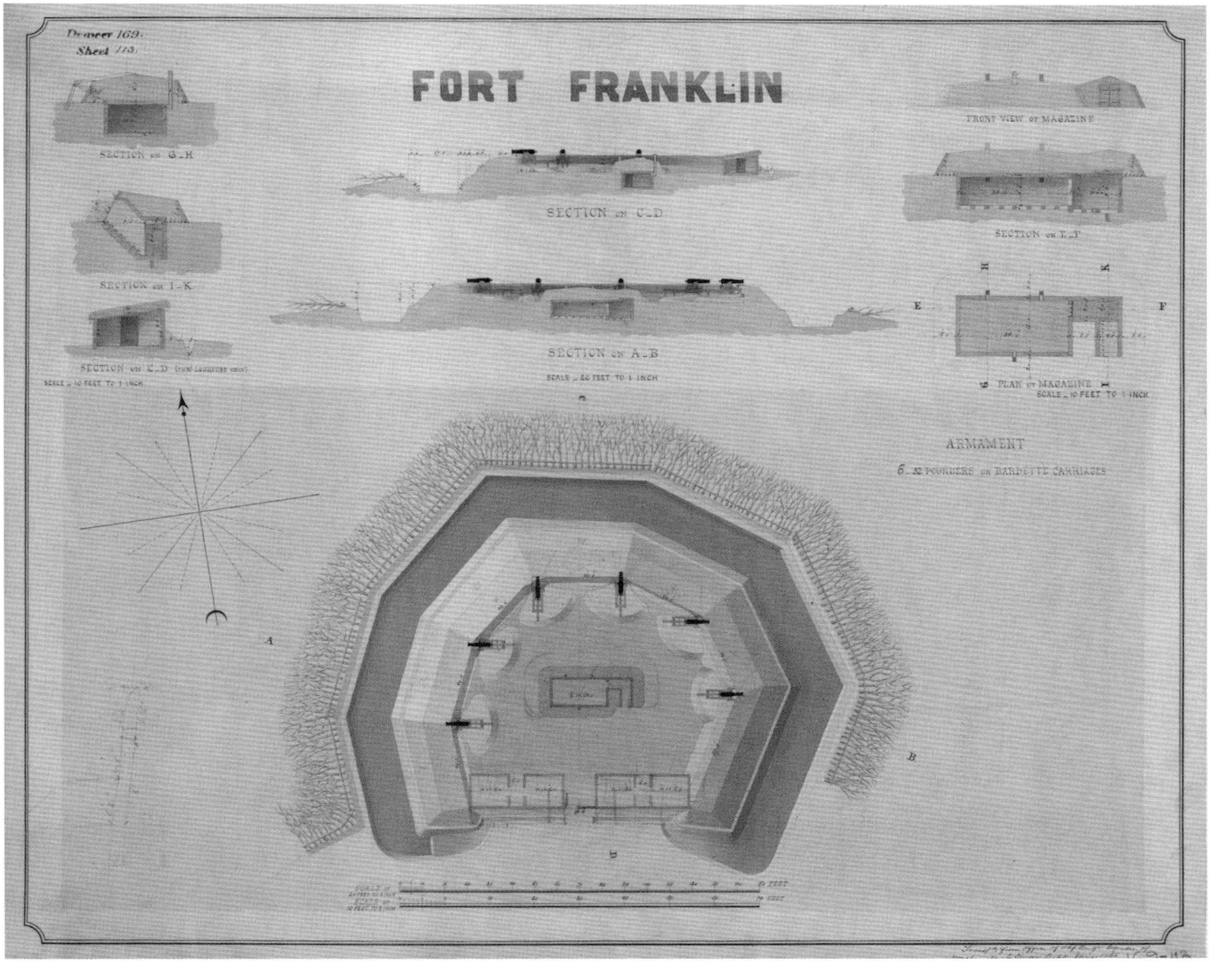

Plan and sections of Fort Franklin. (Courtesy of New York Public Library (CC0 1.0))

Oliver De Lancey was born in New York City and first joined Sir William Howe on Staten Island in 1776. He quickly raised and outfitted three battalions called De Lancey's Brigade. The brigade consisted of 1,500 Loyalist volunteers, primarily from the province of New York. His personal property was plundered in November of 1777 by the Americans and confiscated in October of 1779. Four years later in 1783, De Lancey left New York for England where he died on October 27, 1785, in Beverly, Yorkshire. His grave and a memorial are in Beverly Minister.

General Oliver De Lancey. (Courtesy of Library of Congress (CC0 1.0))

Muskets and rifles

During the Revolutionary War, British infantrymen were armed with several kinds of rifles. Many were issued the Pattern 1776 with a .62 caliber and a 30.5-inch (77.4-centimeter) barrel length. The breech-loading Ferguson rifle had a range of about 100 yards (91 meters) but was one of the few rifles at the time that could be reloaded from the prone position, thus greatly reducing the rifleman's exposure. Also used was the Brown Bess, a flintlock musket that could fire an imposing three to four rounds per minute. By way of contrast, American muskets were manufactured by diverse gunsmiths but because they could be confiscated by the British authorities, many of these rifles did not have a maker's mark. Long rifles that featured grooved barrels, known as Pennsylvania rifles because they were made in the state of the same name by German gunsmiths, were utilized in copious numbers during the war. Although these rifles, that were based on the German Jäger rifle, had an accurate range of 300 yards (274 meters) compared to only 100 yards for smooth-bore rifles, they were complicated to reload, thus reducing their rate of fire. Also to their detriment, they could not be fitted with a bayonet that regular British infantrymen were issued as part of their regular kit. These bayonets were triangular and created deep puncture wounds that in many instances became infected, resulting in death. Because some fighting was hand to hand, the American infantryman was clearly handicapped. The British in many instances favored bayonet charges.

Benjamin Tallmadge, seen here in later life, was one of Washington's finest officers. (G. Parker, courtesy of Library of Congress (CC0 1.0))

The element of surprise had vanished. In April 1781 Tallmadge pleaded with General Washington to allow him to lead yet another attempt to conquer Fort Franklin. Acknowledging Tallmadge's irrefutable abilities as a commander, Washington ordered Tallmadge to create a semiautonomous force consisting of infantry and horsemen with the following caveat: the planned attack would not proceed if the British fleet was anywhere near the fort. Tallmadge was ordered to Newport, Rhode Island, after he had verified enemy strength and positions on Lloyd's Neck. Following that, he was to organize French naval backing that was based in Newport.

A stealthy reconnaissance revealed that Fort Franklin's garrison numbered around 800 men, of whom only 500 were correctly armed. The fort was also only guarded by one 16-gun ship, a galley, and two small frigates. If, and it was a big if, Tallmadge could enlist French support, he reckoned he could take the fort and its smaller sister Fort Salonga, situated eight miles to the east. Although Comte de Rochambeau, commander of the French Army, and Le Chevalier Destouches, commander of the French Navy, were sympathetic, the ships needed by Tallmadge were unavailable.

Jean-Baptiste-Donatien de Vimeur, Comte de Rochambeau, Marechal de France (1725–1807), by Charles-Philippe Larivière. (PHGCOM courtesy of Palace of Versailles)

Comte de Rochambeau: Jean-Baptiste Donatien de Vimeur, better known as Comte de Rochambeau, was born in 1725 and by 1747 was a colonel having served in Bohemia, Bavaria, and on the River Rhine during the War of the Austrian Succession, where he participated in the siege of Maastrict. During the Seven Years' War he fought at the battles of Minorca, Krefeld, Corbach, and Kloster Kampen. By 1780 he was a lieutenant-general commanding 7,000 troops of the Expédition Particulière that joined the Continental Army. He had severe misgivings about the expedition but nevertheless he arrived at Newport, Rhode Island, on July 10 where he remained for an entire year because he didn't want to abandon the French fleet that was being blockaded by the British in Narragansett Bay. He and Washington combined forces at the siege of Yorktown and the battle of the Chesapeake. In September 1781 the Marquis de Lafayette added his sizable number of troops and forced Lord Cornwallis to surrender on October 19.

For his service to the Revolutionary War cause, Rochambeau subsequently received many honors and accolades in the United States. The French named an ironclad frigate after him, as did the United States Navy during World War II: a transport ship, the USS *Rochambeau*. The Washington–Rochambeau Revolutionary Route was designated in 2009 in the Omnibus Public Land Management Act as a National Historic Trail. Rochambeau House is a mansion on the Brown University campus. There is a Rochambeau Middle School in Southbury, Connecticut, along with many streets and shopping centers in the state. The Lycée Français (French high school) in Bethesda, Maryland, is named after him. There is an avenue in his honor in Providence, Rhode Island, and the Bronx, New York, along with a plaza in Springfield, Virginia. Also, in New York, on the Historic Guard Hill in Bedford Corners a farm is named after him, as is a playground in the Richmond District in San Francisco, California. At French Hill in Marion, Connecticut, is a monument to him, and there is a statue of him on Pennsylvania Avenue in Washington, D.C. Also in Washington is a bridge over the Potomac River that bears his name.

Once again, the Americans had been thwarted by the Tory citadel. But in June, after the British attacked near Guilford and captured a 12-oared gunboat near Setauket, Long Island, the French had a change of heart. A small French fleet of eight ships that carried a mixed force of 450 French and Americans led by Baron Anglesey entered Huntington Harbor on July 10 and landed on the eastern side of Lloyd's Neck at 8 a.m., whence they began the two-mile trek to the fort.

The fort today:

The fort no longer exists and there is no dedicated website for Fort Franklin. Its site, with panoramic views of Long Island Sound, Cold Spring Harbor, and Oyster Bay, subsequently became the site of a huge, luxurious residence named Fort Hill House.

Lieutenant Joshua Upham, who commanded the fort, waited patiently until the attackers were around 100 yards from the fort and, critically, were in open ground with no protection. The attackers had walked into a large killing field in which there was no cover. Then, Upham hit the attackers with a withering barrage of grapeshot fired from two 12-pounder cannons. The initial volley, although it resulted in only three casualties, dissuaded the French, who retreated to their waiting ships and sailed up Long Island Sound out of harm's way. At the same time, riflemen peppered the Americans with enthusiastic but ineffectual musket fire. Alone, the Americans had no choice but to retreat as well. The Loyalists suffered no casualties.

Charles René Dominique Sochet, Chevalier Destouches, was a *chef d'escadre* or admiral in the French Navy. He was born in 1727 and by 1767 was a captain. Subsequent to France entering the American Revolutionary War, he commanded the 74-gun *Neptune*. This behemoth was a ship of the line; her gun deck was 195 feet (60 meters) long, while her beam was 51.4 feet (16 meters) long. Her complement was 690 officers and men. The *Neptune* was armed to the teeth with 30 x 36-pounders, 32 x 24-pounders, 18 x 12-pounders, six of which were on the forecastle, and six 36-pound howitzers.

Admiral Charles René Dominique Sochet. (Wikimedia, (CC0 1.0))

Fort Franklin was never conquered by American force of arms. Its demise came about only when Britain accepted American self-government. The fort was forsaken, signalling the end of British military installations in Suffolk County. At the end of the war, it became a refugee camp for Loyalists and in 1783 it was demolished. Local inhabitants enthusiastically scavenged anything of value left behind by the British.

Fort Golgotha

Lieutenant-Colonel Benjamin Thompson (later Count Rumford) ordered the building of this fort in 1782 after he had destroyed Huntington, New York's burial ground and church. Thompson commanded the King's American Dragoons, who were mounted infantry created from a miscellany of other British Loyalist units, including black soldiers. Thompson's men saw service primarily on Long Island in 1782 and early 1783. In July 1783, he and his soldiers were transferred to Saint John, New Brunswick. They were subsequently disbanded in October. Thompson never saw combat and is only remembered for his macabre—or to put it charitably, poor taste—decision on where to build Fort Golgotha.

Born in Massachusetts, Thompson had applied to join General Washington's staff but was rejected. Insulted and angry, he switched his allegiance and joined the British who readily accepted him. It took Thompson a relatively short 15 days to build the fort, which was part of a network of forts around Huntington. To the

Old Burial Hill, Fort Golgotha. (Olivia Mobbs courtesy Flickr (CC BY 2.0))

Lieutenant-Colonel Sir Benjamin Thompson, Count von Rumford. (Courtesy Smithsonian Institute (CC0 1.0) {{PD-US-expired}})

east of Huntington was a larger fort, then known as Gallows Hill and now known as Fort Hill. Fort Salonga lay farther to the east, while Fort Franklin was to the north on Lloyd's Neck.

Thompson's enmity towards the Americans who had rejected him expressed itself in many bizarre and ghoulish forms, including latent sadism and other manifestations that were in all likelihood clinically diagnosable psychosis. This is what he did: after the Old Fort Presbyterian Church was dismantled, he used the good timbers to construct the fort, thus saving time and energy. He then forced, probably at either bayonet or musket point, the town's residents to build the fort and other buildings in the vicinity. In what was perhaps the most grotesque act of all, he removed over 100 tombstones from the burial ground and used them for the fort's ovens, fireplaces, and foundations of floors. Legend has it that on the bottom crust of loaves of bread baked in the ovens was the reverse inscription that appeared on the headstones of local residents' loved ones. Thompson also ensured that the grave of Reverend Ebenezer Prime, the Old First Church's third minister and a supporter of the Revolution, be situated at the fort's exit so he could walk over it every time he left the fort.

The British left the fort in March of 1783 after occupying it for only four months. In a final act of wanton and needless cruelty, Thompson burned all the wood in the surrounding area to deprive the residents of heating and cooking fuel.

The fort today:

The cemetery is in Huntington, New York, but Fort Golgotha no longer exists. It was, however, added to the National Register of Historic Places in 1981.

Fort Herkimer

This star-shaped fort was situated on the southern side of the Mohawk River, across from the mouth of its tributary West Canada Creek in German Flatts. In 1740, the Johann Yost Herkimer (Herscheimer) family had a homestead in the area and around it they built the first, more rudimentary iteration of Fort Herkimer. Johann was the father of Nicholas Herkimer. The fort was originally named Fort *Kaouri*, or Fort Bear.

More of a house than an actual fort, it was two stories high, 40 feet (12 meters) wide, 70 feet (21 meters) tall, and featured gun loopholes for muskets on every floor. The house was laboriously constructed of 2-foot-thick local stone. The dwelling and abutting defenses were further fortified by Sir William Johnson and it was then that the fort became formally known as Fort Herkimer. When German Flatts was attacked by the French in 1757 and 1758, Nicholas Herkimer, then a captain, had his first experience as leader of colonial forces.

When enemy incursions became more frequent and predictable during the drawn-out affair that was the French and Indian War and crept into the Mohawk River Valley, Fort Herkimer was undone, dismantled, and moved piece by piece by grunt labor to the site of the Herkimer Dutch Reformed Church that was surrounded by a stout log wall and armed with a single swivel gun strategically placed at the top of the church

Plan of Fort Herkimer, at German Flatts, 1756. (Courtesy of William L. Clements Library)

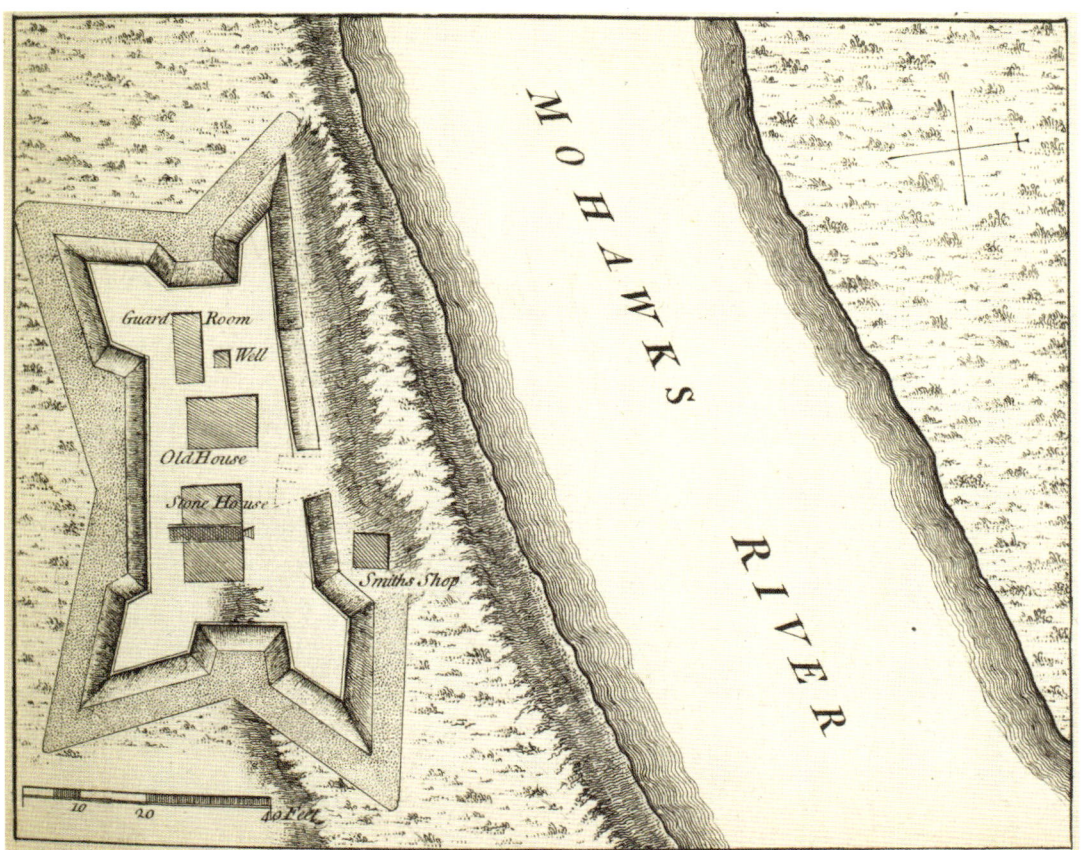

77

Fort Herkimer Church, photographed by Nelson E. Baldwin, December 1936. (Courtesy Library of Congress (CC0 1.0))

steeple. Although usually found on ships, swivel guns when compared to cannons were obviously smaller and much easier to move. To an extent, they functioned like giant shotguns. Earthworks and a palisade provided additional reinforcement and no doubt gave the townspeople a greater sense of security. This church was in turn absorbed into the fort and its name became the Herkimer Church Fort. On the other side of the Mohawk River a blockhouse was built by the British in 1756 and was named Fort Herkimer.

The relatively spacious British fort was by comparison well equipped and featured a parade ground, barracks, guardroom, a separate kitchen, a mess room for officers, a well that provided more than enough fresh water, a drawbridge, a large oven, four sentry boxes on each corner where guards were stationed 24 hours a day, a smith's shop, a terrace, palisades, a parapet, and was minimally armed with 10 swivel guns. Although lightly armed, the fort was more than capable of defending itself as its garrison also included marksmen who were proficient in musketry.

After the terroristic raid by the nefarious and justifiably feared Indian leader Joseph Brant, approximately 700 people found themselves homeless. Another raid by 600 Indians and Tories in the same area—German Flatts—in 1780 yielded almost identical results. Many sought refuge in nearby Fort Dayton. At the end of the war the fort, like many others, was abandoned in 1783. However, in February, during the middle of a frigid winter in which the temperature plunged to mind-numbing lows, Colonel Marinus Willett took it upon himself to sortie out from the relatively cozy confines of the fort to conquer Fort Ontario in Oswego. Overcome primarily by the debilitating effects of the arctic-like temperatures, Colonel Willett wisely returned to the safety and warmth of Fort Herkimer. The fort subsequently became an important and much-needed supply depot for forts to the west, including Fort Niagara and Fort Detroit.

The fort today:

In 1812 the church was refurbished and expanded. It still exists. In 1918, due to the building of the Erie Barge Canal, the ramparts were destroyed. Confusingly, the village of Herkimer was developed not on the site of Fort Herkimer but on the site of Fort Dayton.

www.visitfortherkimer.com is one of several websites devoted to Fort Herkimer and the Herkimer Church Fort.

Fort Jay

Governors Island is in New York Harbor in the borough of Manhattan. It is 172 acres in extent and is about 800 yards (730 meters) south of Manhattan Island. Brooklyn lies to the east, separated by the 400-yard-wide (370-meter-wide) Buttermilk Channel. In 1966 the island was transferred to the United States Coast Guard and it became home to the Atlantic Area Command, the Maintenance and Logistics Command, and the Captain of the Port of New York. Coast Guard cutters celebrating the World War II Allied invasion of Normandy left from Governors Island. The Coast Guard left in 1996 due to budget constraints.

The original name in the native Lenape language was Paggank, or Nut Island, so named because of the proliferation of oak, chestnut, and hickory trees. In 1524 Giovanni da Verrazano was the first European to lay eyes on the island. In May 1624 Adriaen Block, a Dutch explorer, named it Noten Eylandt and it was the first place that settlers from the Dutch Republic landed from their ship *New Netherland*, commanded by Cornelius Jacobson May. Thirty families disembarked and that is why the New York State Senate and Assembly recognize the island as the birthplace of New York State. The name was later anglicized to Nutten Island.

During the colonial period, the British colonial assembly reserved the island for exclusive use by New York's royal governors. In 1774 the apostrophe between the 'r' and 's' was dropped. The island was used as a preserve to breed and hunt pheasants by one governor, while others leased out the land for profit. Around 1710 the island was briefly used as a quarantine station for refugees but it remained for the most part relatively pristine—until the Revolutionary War and the years that preceded it.

In expectation of war with France plans for defenses on the island were made in 1741 but the forts were never built. In 1775, during the French and Indian War, the island was utilized as a military bivouac when Sir William Pepperell, who commanded the 51st Regiment of Foot, led his troops onto the island. Approximately 10 years later, other regiments of British troops were on the island and a crude fort was built in conjunction with protective earthworks. Although the British military engineer John Montresor promoted plans to improve the existing fortifications, these came to naught.

Realizing the strategic importance of New York Harbor, Washington tasked General Charles Lee to craft a defensive plan for the harbor. This all-inclusive plan included forts in Brooklyn, the

Aerial view of Fort Jay, taken in 1982/3. (Library of Congress, Prints & Photographs Division, NY,31-GOVI,1-11 (CC0 1.0))

A 10-inch Rodman at Fort Jay. (Jet Lowe courtesy of Library of Congress, Prints & Photographs Division (CC0 1.0))

Battery, and on Governors Island. Enter the man whose energy knew no limits: Israel Putnam was responsible for building the star-shaped Fort Jay and arming it with eight cannons.

Fort Jay is the oldest defensive structure on Governors Island. It was named after John Jay, who was a member of the Federalist Party, a Chief Justice of the Supreme Court, a Secretary of State, and one of the founders of the United States. More guns were later added, bringing the total to around forty. The idea of Fort Jay and the other forts in the area was to give the British pause before they sailed uncontested into the harbor. Fort Jay's guns had a modest success when they engaged HMS *Phoenix* and HMS *Rose* as they ran past on their way to Tappan Zee. However, the British commanders were very much aware of the damage Fort Jay's cannons could inflict, which in turn made them wary about entering the East River. The forts in New York Harbor were also instrumental in enabling General Washington to retreat from Brooklyn to Manhattan after the battle of Long Island, when British troops attempted to take Brooklyn Heights in what was the largest battle of the war on August 27, 1776. The British ships that were two miles (3.2 km) downriver suffered no damage from the harbor's collective guns.

In a series of strategic moves and countermoves the British then deployed to Manhattan, the Continental Army vacated both Brooklyn and Governors Island, and the British moved back onto Governors Island. All of this precipitated a long cannon duel that began on September 2 and ended on September 14 between the British Governors Island guns and General Washington's guns on the Battery that was located in front of Fort George in Manhattan. Midway through the duel there occurred one of the most innovative events of the war, involving the United States submersible *Turtle* in what was the first known submarine attack in history.

The *Turtle* looked more like a clam standing on end than either a turtle or terrapin. It was 9 feet 10 inches long (3 meters), had a beam of 2 feet 11 inches (0.9 meters), and a propulsion speed in calm waters by means of hand-cranked propellers of 3 mph (4.8 kph or 2.6 knots). It was made of oak and brass and held together with heavy wrought-iron hoops in much the same way as whiskey barrels are bound. The wooden shells were slathered with tar to increase water tightness. The idea was to affix a bomb or other explosive device to the *Turtle*, sneak up, preferably at night, on an unsuspecting British ship lying at anchor, affix the bomb under the ship's water line and then make a getaway before the bomb exploded. There was enough air in the *Turtle* for only 30 minutes. The operator then had to surface to replenish his air supply via a ventilator. The idea of underwater explosives first came to the American inventor David Bushnell while he was at Yale University and, by 1775, he had actually created a method of detonation that was a clock-driven mechanism attached to a flintlock musket. The submarine had many moving parts made of brass that were built by Isaac Doolittle, a New Haven-based, multitalented engraver, clockmaker, inventor, brass manufacturer, and silversmith. Doolittle's place of business was half a block from Yale University. Doolittle was known for his large and complicated brass-wheel hall clocks. He also invented a mahogany printing press in 1769, the first ever made in America.

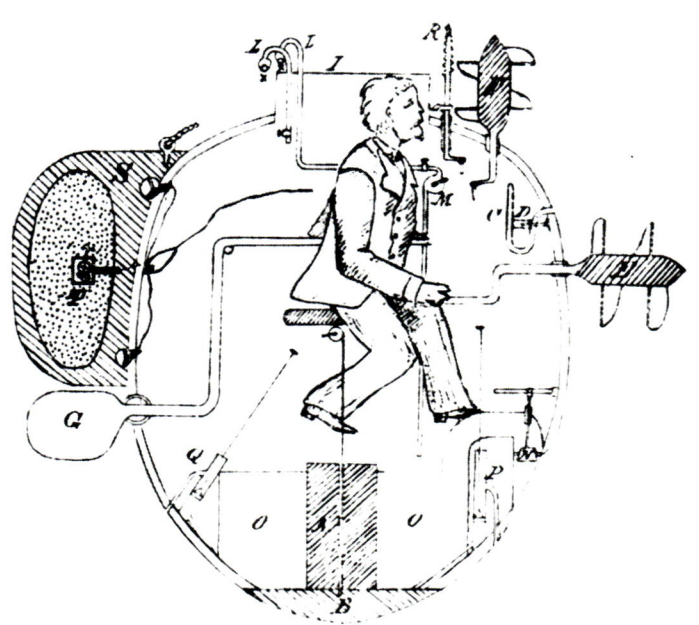

Drawing of a cutaway view of the *Turtle* made by Lieutenant-Commander F. M. Barber in 1885 from a description left by Bushnell. (CC0 1.0)

A model of Bushnell's *Turtle* at the Oceanographic Museum, Monaco. (Kyriaki courtesy of Flickr (CC BY-SA 2.0)

He also made brass compasses, surveying instruments, owned a brass foundry where he cast bells, and when the American Revolution began, the patriotic and wealthy inventor built a gunpowder mill in New Haven and was sent to prospect for lead by the Connecticut government. The Governor of Connecticut, Johnathan Trumbull liked the invention's potential and lobbied General Washington for funds for the development and testing of the submersible.

Israel Putnam has eight counties named after him, eight streets and roads, one Russet apple, two state parks, one pond, and several ships including the USS *General Putnam*.

The irrepressible, ubiquitous Israel Putnam, one of America's finest. (Fabronius courtesy of courtesy of Library of Congress (CC0 1.0))

The propulsion system, or propeller, was simply two oars or paddles used for moving the *Turtle* forwards and backwards. They were 12 inches long (30 centimeters) by 4 inches wide (10 centimeters). The craft also had navigation instruments, brass foot-operated water ballast and forcing pumps, depth gauge, compass, a brass crown hatch, a clockwork detonator for the bomb or mine, a hand-operated propeller crank, and a foot-driven treadle with a flywheel. Natural light was provided by six small sections of thick glass at the top of the *Turtle*. The internal instruments had small bits of bioluminescent foxfire attached to the needles so the operator could see his position in the dark. However, when the temperature dropped too low, the illumination failed, rendering the machine essentially blind. The pilot would have to be in excellent physical condition be able to correctly read instruments and not suffer from claustrophobia. He would have to adjust the bilge to prevent the *Turtle* from sinking and at the same time be working the hand crank that made the propellers in the front of the machine go forward and backward, and also steer the submarine by using a lever that connected to the rudder located in the rear.

Sergeant Ezra Lee. (From *The Story of the Submarine* by Farnham Bishop)

In August 1776 Bushnell asked for volunteers to operate the submarine and eventually, at 11 p.m. on September 6, Sergeant Ezra Lee piloted the *Turtle* toward Admiral Richard Howe's flagship HMS *Eagle* that was anchored off Governors Island. The *Eagle* was indeed a very worthy target and if she could be sunk it would be a tremendous psychological and military blow. Howe's flagship was a proper ship for an admiral. She was a 64-gun, third-rate ship of the line. Her gun deck was 156 feet 6 inches long (48.6 meters), her beam was 44 feet (13.5 meters), and her draught was 10 feet 8 inches (3.25 meters). She boasted 26 x 24-pounders on her gun deck, 26 x 18-pounders on her upper gun deck, 10 x 4-pounders on her quarterdeck, and two 9-pounders on her forecastle.

It took Sergeant Lee two hours fighting a strong current in the darkness to reach his destination. He had imprudently begun his mission with only 20 minutes of air. Nevertheless, he surfaced, lit the fuse on the mine, and tried to stab it into the hull of the *Eagle*. Several attempts failed and time was running short for Lee, who justifiably feared that the resulting explosion might kill him.

Unbeknown to the Americans, the British had begun installing copper sheathing on the hulls of their warships to protect them from damage caused by shipworms and other marine life. Lee made one

The Sally Port at Fort Jay, photographed in 1891. (New York Public Library (CC0 1.0))

The inner works of Fort Jay. (King of Hearts (CC BY-SA 4.0))

more attempt before beating a hasty retreat. It is possible he may have been suffering from exhaustion or carbon monoxide inhalation. The charge floated into the East River where it exploded, sending huge spouts of water into the air. Lee made one more attempt on October 5, when he tried to attach a mine to the hull of a British frigate anchored off Manhattan. He was seen by the ship's watch and had to beat a retreat. The *Turtle* disappeared several days later on board its tender vessel that was sunk by British naval fire near Fort Lee, New Jersey. It was never recovered.

Apparently, Bushnell never recovered either. To his credit, in 1777, he made mines that were to be towed so they could blow up HMS *Cerberus* near New London harbor and floated down the Delaware River to play havoc with the British fleet that was stationed off Philadelphia. Both attempts failed. Bushnell died in obscurity in 1824; however, his innovative ideas were taken to the next level by such noted inventors as Robert Fulton, who slowly transformed the *Turtle* into the *Nautilus*.

In September 1776 the earthwork fortification on Governors Island was abandoned by the Americans, which led to the British occupation of New York City. The existing earthworks on Governors Island were improved by the British and the island was used as a Royal Navy hospital until the British left on November 25, 1783. The island was then transferred to the state of New York. The earthworks gradually deteriorated due to the ravages of time and weather. By 1784 the fort was in ruins and the state of New York decided to make improvements to the unusable fort. This time it was rebuilt as a square with four bastions and named after the Governor of New York, John Jay. Money was appropriated and shifted around from one source to another, while Congress in 1797 allocated funds for continuous rebuilding. The state of New York then shifted responsibility for both Governors Island and Fort Jay to the federal government by selling both entities to the government for $1. In essence, the state of New York had fobbed off its financial responsibility for the island and the fort to the federal government that in turn, recognizing the strategic importance of the island, wasted no time in spending money to improve the fort.

It was a case of out with the old and in with the new. In 1806 the earthworks, such as they were, were immediately replaced by stronger, longer-lasting granite and brick walls, and the fort itself was enlarged under the direction of Major John Williams, the Chief of Engineers of the Army Corps of Engineers. Williams also supervised other fortifications in New York Harbor. (Williams was the first Superintendent of the United States Military Academy and a grandnephew of Benjamin Franklin. He was also elected to the Fourteenth United States Congress.) The 1806 rebuilding consisted of sandstone walls and an arrow-shaped ravelin (a triangular fortification or detached outwork, located in front of the inner works of a fortress), surrounded by a waterless moat. The enlargement enabled the fort to become what was known as the Second System of United States seacoast fortifications.

In front of the moat, trees were cleared to create a glacis or clear, unimpeded field of fire aimed directly at any enemy force foolish or brazen enough to attack the fort. The glacis was also a very effective way of stopping or hindering cannonballs fired from enemy warships. The fort was situated on the highest point of the island. Wood and brick barracks for officers and regular troops were later built. When the rebuild was complete, there was coincidentally a change in the presidential administration and the name of the fort was changed from Fort Jay to Fort Columbus, probably after Christopher but no one seems to know for sure. The name change occurred between December 15, 1806, and July 21, 1807. An anecdote attributed in 1913 to Edmund Banks, who was an Episcopal priest, a United States Army chaplain, and the author of a book on the history of Governors Island, has it that the new Republican administration was displeased with the Jay Treaty between the United States and Britain. The fort's name eventually reverted to Fort Jay in 1904.

Refreshed, rebuilt, and rejuvenated, Fort Columbus née Fort Jay acted with Fort Wood on Liberty Island, Fort Gibson on Ellis Island, Castle Clinton at the Battery in lower Manhattan, and two other forts on Governors Island, Castle Williams and South Battery, to successfully deter any hostile ambitions the British Navy might have had against New York City during the War of 1812. There were easier, less-well-defended targets such as the Great Lakes, the Chesapeake Bay, and the Gulf of Mexico south of New Orleans.

In the early 1820s Fort Lafayette, Fort Richmond, and Fort Hamilton located at the Narrows of New York Harbor lessened the necessity for forts in the Upper Harbor, so the United States Army did the financially responsible thing and sold some Upper New York Bay forts to the state of New York at an impossible-to-refuse price and transferred others to federal agencies.

Governor John Jay, after whom the fort was named. Painted by Gilbert Stuart. (National Gallery of Art (CC0 1.0) {{PD-US-expired}})

Fort Columbus/Jay was the exception. It sat on 68 acres, was only 1,000 yards (914 meters) from the tip of Manhattan, and there was more than enough room to accommodate a modest garrison. It was practical and cost efficient, so it was spared the fate of the other forts. Moreover, the fort was the closest major army post to the United States Military Academy at West Point and served as a convenient and handy first posting or departure point for freshly minted officers who were in transit to other installations on either the Atlantic or Pacific coast. The list included Ulysses S. Grant and Robert E. Lee. Then as now, location was everything.

By the 1830s weapon technology had improved immensely, lessening the defensive worth of the fort but, like weaponry, it continued to evolve, and the army found new uses for the fort. Starting in 1833 the fort got a facelift that took the form of four new barracks. The barracks were in the Greek revival style, unified by two-story Tuscan porticos and occupied by officers and enlisted personnel. Not to be left out, the Ordnance Department built the New York Arsenal adjacent to the fort but not actually part of it. It was used as a distribution center for weapons both manufactured and contracted for posts across the United States. The South Battery became in 1836 the Army School of Musical Practice where young boys were trained to become company drummers, fife players, and regimental musicians. In 1852 the General Recruiting Service was located at the fort.

Plan of Fort Columbus/Jay. (M. Vincent courtesy of Library of Congress Prints and Photographs Division (CC0 1.0))

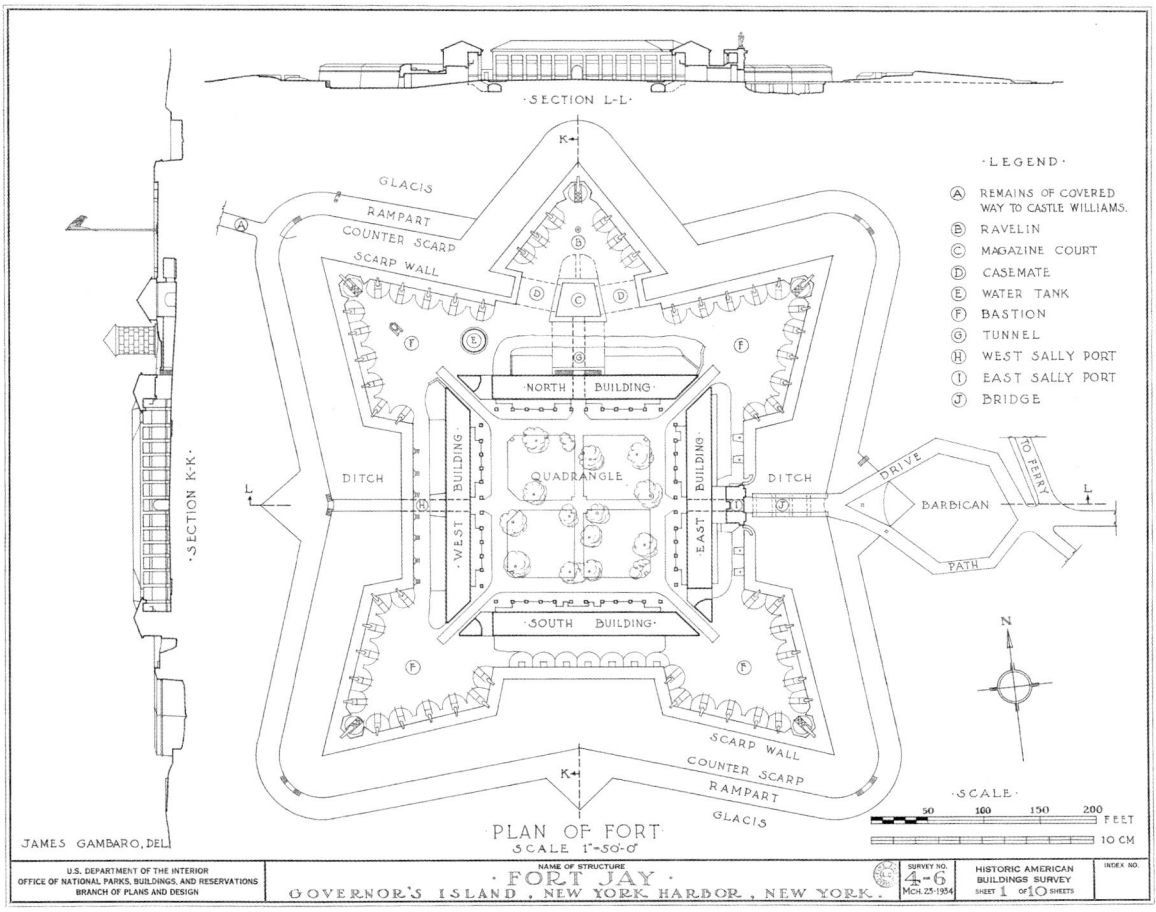

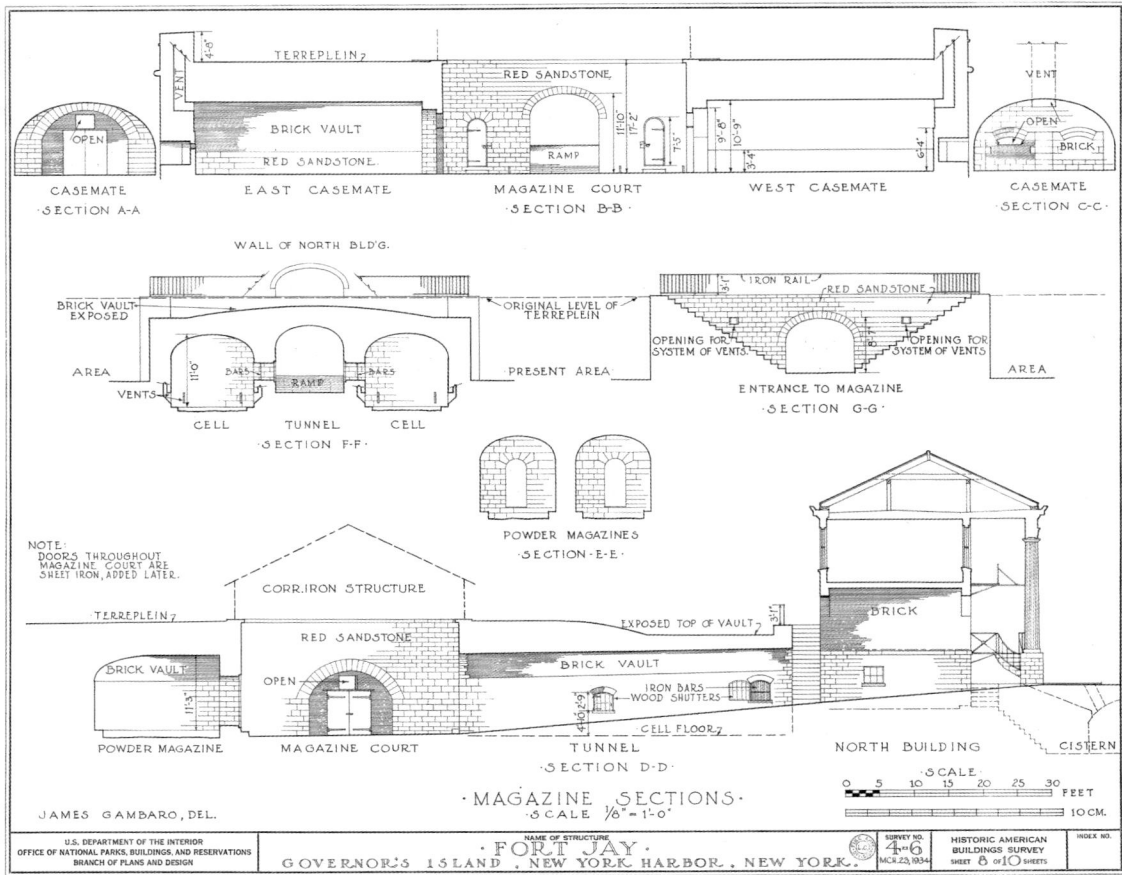

Magazine section, Fort Columbus/Jay. (M. Vincent courtesy of Library of Congress Prints and Photographs Division (CC0 1.0))

Fort Columbus/Jay continued to serve the nation. In December 1860 and April 1861 the army tried to send supplies and men to aid the beleaguered Fort Sumter in Charleston, South Carolina. The first attempt involved the chartered steamship *Star of the West* that was fired on by cadets from the Citadel, South Carolina's military college of. The second attempt also met with failure when new enlistees from the fort prompted South Carolina rebels to attack Fort Sumter, thus starting the Civil War. The north barracks were used to confine Confederate officers before they were sent to other prisons. Following the Civil War, Fort Columbus/Jay got another upgrade in its weapons that took the form of almost 50 x 10-inch (254-millimeter) and 15-inch (381-millimeter) Rodman cannons.

Enter Elihu Root, President Theodore Roosevelt's Secretary of War and an influential lawyer who wanted to make more improvements to Fort Columbus/Jay. He was instrumental in the expansion of Governors Island from 60 acres to 172 acres: using an abundance of readily available landfill from the newly built New York City subway system and harbor dredging. In February 1904 he restored the original name of Fort Jay to the installation.

A scrap drive during World War II left the fort with only four 10-inch and one 15-inch cannons, positioned at the fort's east entrance gate and north ravelin as decorations. Also during the war, Fort Jay was headquarters of the First Army and subsequently the Eastern Defense Command that coordinated all army units and defenses such as coastal defenses, antiaircraft, and fighter plane assets. Beginning in April 1942 United States Army forces in Newfoundland and Bermuda were included in the Eastern Defense Command.

Interior details of East Entrance, Fort Columbus/Jay. (Jet Lowe courtesy of Library of Congress Prints and Photographs Division (CC0 1.0))

Fort Jay's usefulness continued, although in November 1964, after careful consideration of various ways to save money, the U.S. Army decided to close Fort Jay. Seeing an opportunity, the United States Coast Guard moved in and used the fort until September 1966. Fort Jay had hung on like grim death for as long as it possibly could but this tine its demise was final and irrevocable.

Plan of a magazine, Fort Columbus/Jay. (M. Vincent courtesy of Library of Congress Prints and Photographs Division (CC0 1.0))

· PLAN OF MAGAZINE ·
SCALE ⅛"=1'-0"

JAMES GAMBARO, DEL.

U.S. DEPARTMENT OF THE INTERIOR
OFFICE OF NATIONAL PARKS, BUILDINGS, AND RESERVATIONS
BRANCH OF PLANS AND DESIGN

NAME OF STRUCTURE
· FORT JAY ·
· GOVERNOR'S ISLAND , NEW YORK HARBOR , NEW YORK ·

SURVEY NO.
4–6
MGH.Z3-1934

HISTORIC AMERICAN
BUILDINGS SURVEY
SHEET 7 OF 10 SHEETS

INDEX NO.

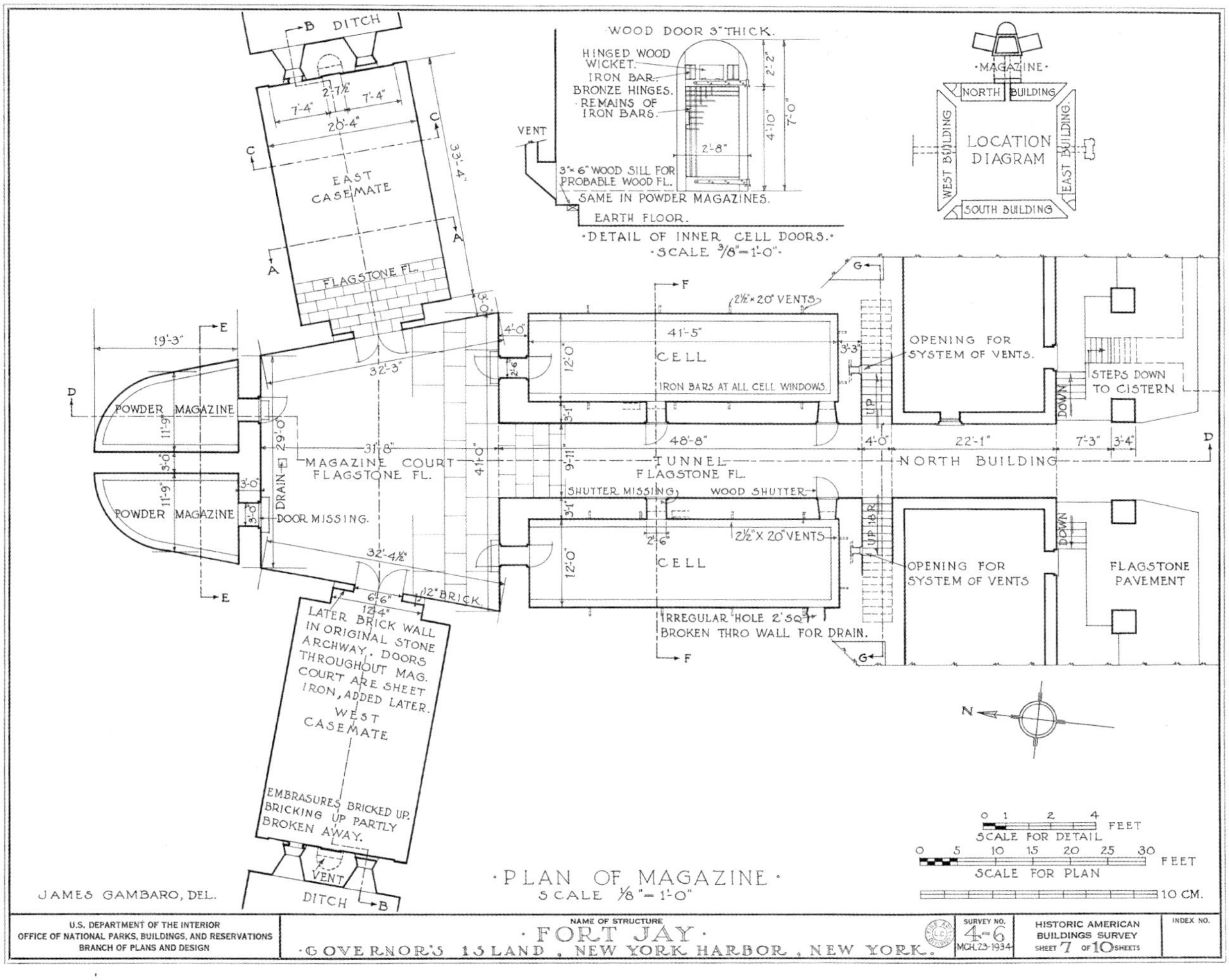

The gate of Castle Williams. (Axel Tschentscher (CC BY-SA 4.0))

Castle Williams, although not a Revolutionary War fort (10 South Street, New York, NY 10004), is a circular defensive fort made of red sandstone that was also used as a prison, and is located on the western point of the island. It's an easy climb to the roof where you'll get breathtaking views of lower Manhattan. The fort was designed and completed by Colonel John Williams during 1807–11. It was the first casemated battery in the United States. Castle Williams was listed in the National Register of Historic Places on July 31, 1972, and, along with Fort Jay, is part of the Governors Island National Monument.

The fort today: The City of New York eventually took over the daily operation of Governors Island, opened it for public use in 2005, and turned it into what is essentially a giant playground for both adults and children that offers a welcome and sometimes much-needed respite from the inherent stresses of daily city life. There are also 52 historic buildings on the island. The National Park Service administers a small portion of the northern end of the island as the Governors Island National Monument and this is where Fort Jay and Castle Williams are sited. The remaining 150 acres are operated by the Trust for Governors Island as a public park. The New York Harbor School, a maritime-focused high school, is also on the island.

Open daily year-round, the island is car free and you can hike, run, ride a bicycle, or kayak. The island hosts free arts and cultural events. The highest elevation on the island is Lookout Hill, which is 70 feet (21 meters) above sea level. Even during the winter months when New York can get quite cold and is subject to street-clogging snowstorms, there are activities on the island such as ice skating, fire pits, and places to buy hot and cold beverages. A variety of stores sell food and produce from around the world. You can even rent a sled when there's enough snow. On Saturdays through April you can bring your dog and take them to a specific off-leash area so they too can play. Or, you can enjoy the view from different waterfront cafés while sipping craft beer or cocktails. You can also glamp on the island.

Ferries leave every day from the Battery Maritime Building on lower Manhattan, 10 South Street, and from Pier 6 in the Brooklyn Bridge Park near Furman Street and Atlantic Avenue, and also the Atlantic Basin in Red Hook near the intersection of Pioneer Street and Conover Street.

www.govisland.com/plan-your-visit

Fort Klock

Located between State Road 5 and the Mohawk River and two miles east of the village of St. Johnsville, Fort Klock was a combination fort and dwelling or basically a fortified home. It was built by Johannes Klock, a Palatine German, around 1750 and is an excellent example of a typical fort/house that served as a trading post during the French and Indian War and that was later involved in the Revolutionary War. Klock obviously knew what he was doing. The structure is a single-story house built partly on bedrock and locally sourced stone that makes up the foundation. The walls were 2 feet (.61 meters) thick and had many loopholes through which riflemen could fire their muskets. There were two separate rooms on the main floor, with sleeping quarters located in the attic. On the eastern side of the house was a door that led to a basement which was also divided into two chambers separated by a heavy stone wall. Perhaps most importantly, one of these chambers held a utile but small spring-fed pool. In 1776 Klock made another defensive addition in the form of a stockade that surrounded the house/fort. Klock, who was born in Stone Arabia, Montgomery County, New York, served in the 2nd Regiment of the Tryon County Militia, New York.

The Tryon County Militia was authorized on March 8, 1772, when the Province of New York enabled the formation of systematized militia for each county in the colony. By 1776 the Tryon County Militia had in fact become a functioning, well-organized military force under the auspices of the Tryon County Commission of Safety. It was a well-led unit of veteran troops whose collective resumé spoke for itself. The militia participated in the battle of Oriskany, the attack on German Flatts, the battle of Cherry Valley, the battle of Klock's Field,

Look toward Fort Klock over the foundations of the old barn. (Library of Congress Prints and Photographs Division (CC0 1.0))

Fort Klock as photographed by Stanley P. Mixon, June 15, 1940. (Library of Congress Prints and Photographs Division (CC0 1.0))

and the battle of Johnstown. There were four regiments in the militia and each regiment was structured to its specific geographic locality. The 1st Regiment was responsible for the Canajoharie District, the 2nd Regiment for the Palatine District, the 3rd Regiment for the Mohawk District, and the 4th Regiment for the German Flatts and Kingsland districts.

The area that surrounded Fort Klock was subject to numerous raids by Native American Indians who were usually led by British officers. These raids destroyed houses, barns, and crops and were a source of major concern among the early settlers. It was here in a field just to the west of Fort Klock, on October 19, 1780, that a fight known as the battle of Klock's Field, sometimes referred to as the battle of Failing's Orchard, the battle of Nellis Flats, or the more colorful battle of Stone Arabia, took place.

The opponents were Lieutenant-Colonel Sir John Johnson 2nd Baronet, who was a wealthy landowner, a Loyalist leader, and a magistrate in Canada. After moving to the United States, he commanded the King's Royal

Fort Klock, photographed in 1991. (Lvklock (CC BY-SA 4.0))

Sir John Johnson, 2nd Baronet. With his ally, Joseph Brant, Johnson was the scourge of the New York frontier. (https://archive.org/ (CC0 1.0) {{PD-US-expired}})

Regiment of New York and was made a brigadier-general. (In 1771 he became the last Provincial Grand Master of Masons in the colonies of the province of New York, New Jersey, and Pennsylvania.) Johnson's force consisted of elements of the 8th Regiment of Foot, the 34th Regiment of Foot, the King's Royal Rangers, Butler's Rangers, and the notorious Joseph Brant's eager-to-take-scalps volunteers. Their mission was to seek and destroy as many American buildings and crops as possible.

Robert Van Rensselaer was a brigadier-general who on June 16, 1780, commanded the 2nd Brigade of the Albany County Militia. Prior to his promotion he had been a colonel of the 8th Albany County Regiment. The 2nd Brigade included the Tryon County Militia. He had previously fought at the siege of Fort Ticonderoga. He had done his best to promote peace by arranging a conciliation with Oneida Nation chiefs who had formed an alliance with the British.

American regiments had between eight and 10 companies. A company had a captain, two lieutenants, four sergeants, four corporals, a fifer, a drummer, and 78 privates. Not all companies, or regiments for that matter, were always at full strength. Men got sick while others were convalescing from wounds and therefore not fit for duty. Drums kept the cadence of large numbers of men marching in unison, while fifers lifted their spirits and performed a similar function as bagpipes did in later conflicts. Both the Americans and British used drummers and fifers. A quarter of each company, approximately 22 men, were selected as minutemen. These were elite troops that were required to be highly mobile and could muster quickly. They were chosen for their enthusiasm and physical strength. In modern terminology they were force multipliers.

Sculpted by Henry Hudson Kitson and erected in 1900, this statue in Lexington, Massachusetts, is commonly called "The Lexington Minuteman." (Daderot (CC BY-SA 3.0))

Robert Van Rensselaer's grave. (Added by Liz on www.findagrave.com/memorial/)

After Sir John Johnson and Brant began razing houses and farms in Stone Arabia, a small community a mile north of Fort Keyser, on October 19 a unit of Massachusetts troops reinforced by New York militia and rangers led by Colonel John Brown sallied out from their post at Fort Paris in Stone Arabia, intent on defeating one of Johnson's detachments. Brown was acting on information provided by deserters who told him that Johnson's force was smaller than Brown's garrison of 360 troops. Brown was met by Johnson's main force which however proved much larger. It proceeded to outflank and encircle Brown's men and defeated them, killing Brown in the process.

Later in the day, Brigadier-General Robert Van Rensselaer led units from Albany County and Tryon County plus New York State levies commanded by Colonels John Harper and Lewis DuBois and caught up with Johnson and his troops at Klock's farm. The fighting was intense and lasted until the British were in turn outflanked by the right column of Van Rensselaer's troops. Eventually, the left column was also successful in outflanking the British, so rather than risk having his men fire on one another and because his men had been marching and fighting continuously for almost 26 hours, Van Rensselaer gave the order to withdraw several miles to the east.

The fort today:

Fort Klock was declared a National Historic Landmark in 1972, and was added to the National Register of Historic Places. The fort's address is 7203 Route 5, approximately two miles (3.2 km) east of the village of St. Johnsville, NY. The fort is part of a 30-acre site that also includes not only the fort but a renovated colonial-era Dutch barn, a blacksmith shop, and a 19th-century schoolhouse. There are reenactments and artifacts plus different celebrations that make a visit to Fort Klock a rewarding, fun, and educational experience.

www.fortklockrestoration.org.

Because they saw how the battle was going, Sir John Johnson, Colonel John Butler, and Joseph Brant and his Indian compatriots hightailed it south of the river, essentially deserting their men and leaving them to fend for themselves. During the ensuing rout, Johnson's men left behind their baggage, prisoners who had been captured earlier, and their cannon. In spite of this defeat, Johnson's campaign was regarded as a major success that resulted in a 20-mile swathe of the Mohawk Valley being laid waste. Van Rensselaer established a camp near the Palatine Church. He was later court-martialed in Albany because he had not pursued the fleeing enemy but was subsequently acquitted.

Fort Montgomery

Fort Montgomery represented one of the first major investments by the Americans on strategic construction projects. If the British seized control of the Hudson River, they could divide the colonies. Preventing them from doing that was paramount.

A plan was hatched. A cable chain supported by a boom would be drawn across the river and tied at one end to Fort Montgomery and to the smaller Fort Clinton at the other end. Fort Clinton had a garrison of 700 men and was on the southern bank of the Popolopen. Fort Montgomery was situated at the confluence of Popolopen Creek and the Hudson River in Orange County near Bear Mountain.

The original site of Fort Montgomery was farther north on the Hudson River, on Martlaer's Island across from West Point. James Clinton and Christopher Tappan, both of whom were very familiar with the area, did the scouting. The fort was to be named Fort Constitution and it would have four bastions. Construction began

Site of the cannon platform at Fort Montgomery overlooking the Hudson River. (Daniel Case (CC BY-SA 3.0))

View of the Hudson at Fort Montgomery, by William Guy Wall. (New York Public Library Digital Collections (CC0 1.0))

in the summer of 1775 and by November 70 cannons were in place. Increased costs and problems involving management and construction plagued the project from the beginning and ultimately it was abandoned. Resources from Fort Constitution were shifted to Fort Montgomery. To protect the chain and the fort was a river battery of six 12-pounders and landward redoubts connected by ramparts situated on a cliff promontory that rose 100 feet (30.4 meters) above the river, giving the gunners a clear field of fire. Any British ship, big or small, that was foolish enough to venture up the river would either have its hull ripped apart by the chain or be blasted to smithereens by the cannons.

At least, that was the plan.

The battle of Fort Montgomery began on October 6, 1777. To reiterate, the British amassed an overwhelming force that consisted of a detachment from the 17th Regiment of Light Dragoons, the 7th Regiment of Foot Royal Fusiliers, the 17th Regiment of Foot, the 26th Regiment of Foot, 52nd Regiment of Foot, 57th Regiment of Foot, 63rd Regiment of Foot, a company from the 1st Battalion 71st Regiment of Foot, plus German forces that consisted of a grenadier company from the 1st Anspach-Beyreuth Regiment and the Regiment von Trumbach (Landgraviate of Hesse-Kassel).

Sir Henry Clinton commanded the British forces. Prior to that, he had been appointed Governor of the Province of Newfoundland, including New York. During the Seven Years' War he was promoted to lieutenant-

PLAN
of the ATTACK of the FORTS
CLINTON & MONTGOMERY,
upon
HUDSONS RIVER
which were Stormed by HIS MAJESTYS FORCES
under the Command of
SIR HENRY CLINTON, K.B.
on the 6th of Octr 1777
Drawn from the Surveys of VERPLANK, HOLLAND & METCALFE.
By JOHN HILLS, Lt 23d Regt
and Asst Engineer.
London, Published by Wm Faden, Geographer to the KING,
Charing Cross, June 1st, 1784.

The British plan of attack on Forts Clinton and Montgomery. (John Hills, William Faden courtesy of Library of Congress Geography & Map Division (CC0 1.0))

colonel. He fought at Bunker Hill and after Major-General William Howe replaced General Thomas Gage Howe named Clinton as his second-in-command.

Opposing them the Americans mustered the Loyal American Regiment, Emmerich's Chasseurs (sharpshooters), New York Volunteers, King's American Regiment, and the King's Orange Rangers. Major-General Israel Putnam, who was based at Peekskill, commanded the Americans.

Emboldened by the arrival in New York City of 1,700 additional troops, General Clinton confidently sailed up the Hudson River with nearly 3,000 troops to engage Fort Montgomery and Fort Clinton. Two frigates, HMS *Tartar* and HMS *Preston*, plus the brig HMS *Diligent*, the galleys HMS *Dependence*, *Crane*, and *Spitfire*, and sloops HMS *Hotham* and *Friendship* were also added to provide additional firepower from the Hudson River.

The sixth-rated frigate HMS *Tartar* had captured many French ships during the Seven Years' War and had a well-deserved reputation as a fast vessel when employed in the English Channel. She was 117 feet, 10 inches long (36 meters) on her gun deck and 96 feet 11 inches long (30 meters) at her keel. Her beam was a spacious 33 feet 9 inches (10 meters). Crewed by 200 officers and men, her upper deck was armed with 24 x 9-pounder cannons and her quarterdeck had four 3-pounder guns plus 12 swivel guns. Prior to the attack on Fort Montgomery,

A German map showing Forts Clinton, Montgomery, and Constitution (bottom right). (Johann Martin Will courtesy of Library of Congress Geography & Map Division (CC0 1.0))

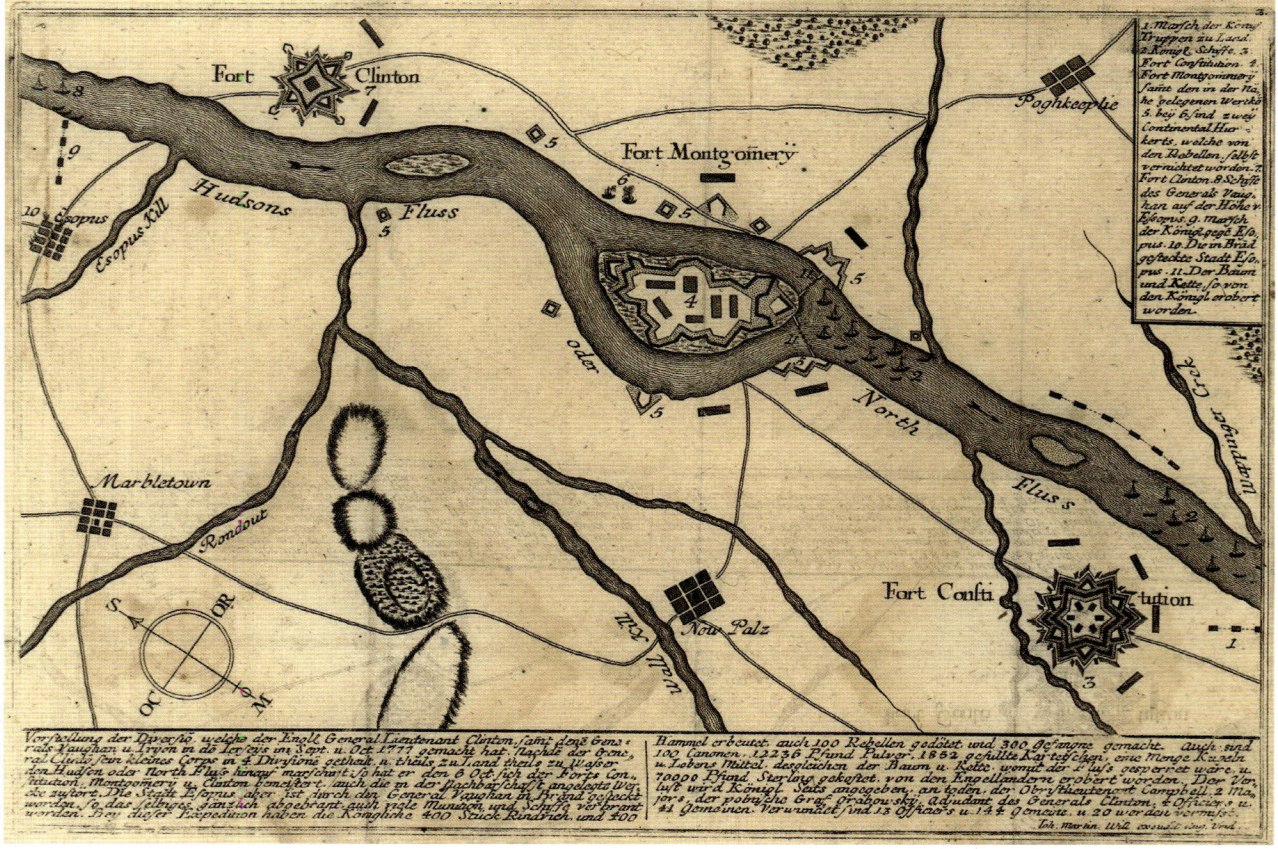

A Hessian private, Erb Prinz Fusileer Regiment of Hesse-Cassel, by Charles M. Lefferts. (Courtesy of New-York Historical Society (CC0 1.0) {{PD-US-expired}})

she had destroyed an American ship off the coast of New Jersey and subsequent to the attack, she captured the Spanish *Santa Margarita* off Cape Finisterre on November 11, 1779. She met her end in April 1797 when she was wrecked off the coast of Saint-Domingue. The fourth-rated ship of the line HMS *Preston* was even bigger and had more cannons. Her captain was John Robinson. Her gun deck was 150 feet-long (46 meters) and her beam was 42 feet 8 inches wide (13 meters). Her gun deck had 22 x 24-pounders, her upper deck had 22 x 12-pounders, her quarterdeck had four 6-pounders, and her forecastle had two 6-pounders. These two ships, together with the others that constituted the British fleet, would play a vital and decisive role in the attack.

It was foggy on the morning of October 6 when Sir Henry Clinton's force of approximately 2,100 men landed at Stony Point. Then, at Doodletown, they encountered a reconnaissance patrol sent out by Governor Clinton. The patrol prudently retreated in the direction of the smaller Fort Clinton. Sir Henry divided his force and attacked both Fort Clinton and Fort Montgomery. Both attacks were from the landward side where the defenses had not yet been completed.

Lieutenant Campbell led the 52nd and 57th Regiments, the Hessian sharpshooters, and approximately 400 Loyalists commanded by Beverly Robinson, seven miles from a gorge toward Fort Montgomery. Sir Henry waited to attack Fort Clinton. The detachment from Fort Montgomery under Captain John Land had only about 100 men in total. They had one small cannon. Retreating a mile from the fort, they clashed with Campbell's men before retreating after spiking their cannon. They made another stand closer to the fort, helped by a 12-pounder,

Cariacture of a Hessian grenadier with a heavy knapsack, dated 1778. ("A Hessian Grenadeir," from the Anne S. K. Brown Military Collection)

A Hessian Grenideir.

but were forced to retreat again. Governor Clinton was offered the chance to surrender but refused. Meanwhile, the British ships were effectively pounding the forts. Return fire was spasmodic and ineffective.

It only took one day to capture Fort Montgomery. Both forts were burned and razed down to their stonework structures. The combined British land and river forces had simply been too much for the Americans. Outmanned and outgunned, they stood little chance of defending their forts.

The fort today:

On November 28, 1972, Fort Montgomery was added to the National Register of Historic Places and was designated a National Historic Landmark. This is a must-see destination that offers something for everyone. Address 690 Route 9W, Fort Montgomery, NY 10922. Access to Fort Montgomery is free to visitors during regular hours. Donations are welcome. There are some program fees for tours and evening events are charged. The fort is compliant with the Americans with Disabilities Act.

The museum contains original items excavated from the fort such as clay pipe fragments, bone-handled forks, pins, Jew's harps, coins, buckles, spoons, cufflinks, buttons, and animal bones that tell the story of the hundreds of soldiers, merchants, servants, laborers, and slaves who lived and worked at the fort. There is also a link from the chain that spanned the Hudson River. It was raised by fishermen in 1861. There are food vessels such as porcelain and salt-glazed stoneware items, case bottles, buff earthenware platters and bowls, and a number of posset pots. There is also a large-scale model of the fort that allows visitors to familiarize themselves with the fort and environs. One of the most crowd-pleasing exhibits are the highly detailed mannequins that depict combatants in battle. A British sergeant in the 52nd Regiment of Foot is depicted loading his lightweight fusil, a musket carried by officers and sergeants. He wears a red regimental coat with buff facings and white buttonhole lace, buff waistcoat and breeches, and black, short gaiters over his shoes. His hat is black and there is a sash around his waist and a short sword at his waist. A private in the Loyal American Regiment attached to Emmerich's Chasseurs, or sharpshooters, is clothed in a green regimental coat with white facings, white wool breeches and waistcoat, and a black cocked hat trimmed with white tape. He has shortened his non-regulation coat so he can move more easily through rough terrain. He is taking aim with a rifle, not a musket. Weapons and other ephemera are also displayed, such as gun parts and musket balls, plus clothing and buttons. There are swords, muskets, bayonets, and artillery projectiles that are displayed in close proximity to where they were actually used. The artifacts, diagram of the fort, mannequins, and weapons create an almost immersive experience that can be enjoyed by everyone.

https://parks.ny.gov/historic-sites/fortmontgomery.com

General Burgoyne's surrender at Saratoga, a painting by John Trumbull. (Courtesy of Yale University Art Gallery (CC0 1.0) {{PD-US-expired}})

The British lost 41 men killed and 142 wounded, while the Americans lost 75 killed and wounded, not counting wounded prisoners. In addition, the Americans were forced to destroy several boats which could not escape upriver due to unfavorable winds.

Captain James Wallace was tasked with clearing obstacles following the battle and reported that the river was clear as far north as Esopus. When he became ill, Sir Henry went to New York and General John Vaughan was put in charge of the forts. Communications by messengers on horses were slow and orders instructing Vaughan to assist General Burgoyne were delayed. It has been speculated, not without good reason, that Burgoyne was unaware of the reinforcements being sent his way. The British victory at Fort Montgomery was Pyrrhic as interruptions caused further delays in reinforcing Burgoyne at Saratoga. At the ensuing battle of Bemis Heights the Americans were too strong for the British and 10 days later forced Burgoyne to surrender at the battle of Saratoga.

Fort Niagara

Although it wasn't designed and built by Americans, Fort Niagara, located on the eastern bank of the Niagara River at its mouth on Lake Ontario, played an important role in the Revolutionary War. The fort's guns controlled access to the Great Lakes and the westward route to the interior of North America. It was a major French military and supply point between the French province of Quebec and their forts in the Ohio country. Indians used the location as a hunting and fishing camp. The French established outposts at the river's mouth in 1679 and 1687. The first fortified structure, called Fort Conti, was built by René-Robert Cavelier in 1678. In 1687 the Governor of New France, the Marquis de Denonville, replaced it with a new fort he named after himself and posted a garrison of

Fort Niagara as seen from Fort George. (From *The Pictorial Field Book of the War of 1812* by Bernard Lossing (CC BY-SA 3.0) {{PD-US-expired}})

FORT NIAGARA, FROM FORT GEORGE.

Aerial view of Fort Niagara. (Larry Koester (CC BY 2.0))

100 men commanded by Captain Pierre de Troyes, Chevalier de Troyes. The Seneca then laid siege to the fort and by the time a relief force arrived from Montréal, disease and severe privation had reduced the garrison to 12 men. In 1744 the French strengthened Fort Niagara and during the French and Indian War huge earth walls and additional buildings were constructed under the leadership of the French engineer Gaspard-Joseph Chaussegros de Léry.

In 1759, the fort was besieged by the British, led by General Jeffery Amherst who took it after 19 days and promptly built a bake house and provisions warehouse. The French sent a relief force to relieve the fort but it was ambushed at the battle of La Belle-Famille on July 24, 1759. Knowing that a relief force was on its way, the British had constructed a breastwork across the road that led to Fort Niagara about two miles south of the fort. Captain Le Marchand de Lignery, who had been preoccupied with organizing an expedition against Fort Pitt from Fort Marchault, received an urgent appeal for reinforcements from Captain Pierre Pouchot. Lignery tasked Charles Philippe Aubry with rescuing Fort Niagara but messengers leaked his plans to the British who seized upon the opportunity to prepare an ambush. Lignery and Aubry walked blithely into the ambush. The trail that ran north from Niagara Falls to Fort Niagara was defended by 464 British regular troops under the command of Lieutenant-Colonel Massey of the 46th Foot. Massey positioned 130 troops from the 46th on his right flank that covered the portage trail in the La Famille clearing. Grenadiers from the 46th and a small number of troops from the 44th Foot were deployed forward from the right flank just above the river gorge. Light companies of the 44th, 46th, and the 4th Battalions of the 60th Foot were positioned to the left of the 46th, while on the extreme left was a contingent from the 44th and New York Regiment. Massey ordered his men to lie down and fix bayonets.

The North Redoubt at Fort Niagara, with British troops (reenactors) falling in. (Ken Smith (CC BY-SA 3.0))

Overlooking Fort Niagara. (Archives of Ontario (Open Government Licence–Ontario))

When the unsuspecting French emerged from the thick woods and came out into the open, the British who were still lying down immediately opened fire and quickly deployed from column to line formation. The French continued to advance. Massey ordered his men to rise and fire another volley. The 46th fired a total of seven volleys before they advanced and fired at will. It was later estimated that each British soldier fired 16 rounds in the fight. While the 46th was advancing, the grenadier company fired in enfilade into the French left flank. Seeing the French begin to wilt under the withering fire, the British charged and fought with their bayonets in close-quarter combat. The French panicked and withdrew. The British chased them for approximately five miles before finally stopping. Lignery was dead and Massey wounded. After he was informed that the expected relief column had been driven off, Pouchot surrendered the fort to Sir William. Johnson.

In the years to come, Fort Niagara would continue to mirror, and be an integral part of, unfolding events. In 1770 and 1771 the garrison strength was 150 officers and regular troops, and stone redoubts were constructed. During the Revolutionary War, the fort was the Loyalist base for Colonel John Butler and his Butler's Rangers, who captured a high-ranking officer of the Continental Army, Colonel William Stacy, in an attack on Cherry Valley, New York. In the summer of 1779 Stacy was held prisoner at the fort where on the riverside flat below the fort crude taverns, bordellos, and stores sprouted, giving Fort Niagara a well-deserved reputation for drunken brawls, whoring, and cheating.

Butler's Ranger, by Charles M. Lefferts. (Courtesy of New-York Historical Society (CC0 1.0) {{PD-US-expired}})

After the 1783 Treaty of Paris ended the War of Independence, the British ceded Fort Niagara to the Americans but the region effectively and practically remained under British control for the next 13 years. Americans occupied the fort in 1796. During the War of 1812, Fort Niagara's cannon sunk the Provincial Marine schooner *Seneca* on November 21, 1812, as she sat in the Niagara River under Navy Hall. The ship was armed with only two cannons and was crewed by approximately 40 men. It had originally belonged to Ebeneezer Hubbard, who had had it seized by the British in Kingston, Ontario.

Once again, the fort changed hands. The British recaptured Fort Niagara on the night of December 19, 1813, in retribution for the Americans burning Newark nine days earlier. The British held the fort until they returned it to the Americans under the terms of the Treaty of Ghent (signed December 24, 1814) that ended the War of 1812.

During World War I, two 90-day officer training camps were hosted at the fort and an elaborate, replicative trench system was constructed to prepare future officers for conditions they would be facing on the Western Front. Throughout World War II, approximately 2,800 German and Austrian prisoners of war were held at the fort which was also used as an induction center. Antiaircraft guns were stationed at the fort from 1952 until the fort was decommissioned in 1963.

An American guard (reenactor) at Fort Niagara. (Florence88 (CC BY-SA 3.0))

The fort today: The fort was designated a National Historic Landmark on October 9, 1960, and added to the National Register of Historic Places on October 15, 1966. There's so much to do and see here that one day might not be enough time to take advantage of all the fort has to offer. The address is 102 Morrow Plaza, P.O. Box 169, Youngstown, NY 14174-0169. The fort is open year-round, but hours are subject to change, so best to check first.

The architectural collection includes six 18th-century buildings that are the oldest structures in the Great Lakes region. The earliest known structure is a French castle that was built in 1726 and is the oldest building in North America between the Appalachian Mountains and the Mississippi River. Other buildings include the above-ground powder magazine (1757), the bake house (1762), provisions storehouse (1762), and two stone redoubts (1770 and 1771). The fort also has three extensive areas that are examples of evolving defensive technologies that spanned a century of military activity in North America. Visitors can explore earthen and log walls and fortified works, stone walls, and artillery emplacements from the first half of the 19th century, and underground brick casemate galleries and outer walls that date from 1862–72 period. The Dauphin Battery has three 6-pounder cannons and two 12-pounder cannons on display.

Visitors will be attracted to the collection that numbers approximately 2,051 artifacts and includes many more cannons plus small arms such as pistols and edged weapons like swords and bayonets. There were also, at last count, 155 pieces of military clothing, 65 accoutrements, 75 pieces of furniture, and military insignia. One of the most spectacular exhibits is the U.S. garrison flag from the War of 1812 that measures 24 (7.3 meters) feet by 28 feet (8.5 meters). The library has 8,000 images of Fort Niagara, 3,000 books and periodicals, and 3,500 manuscript items, plus a large number of reproduced source material curated from other institutions. During special events, reenactors accurately portray and recreate the lives of past garrison members. If you visit during the off season there are self-guided or audio tours, and you can also watch and photograph hourly musket demonstrations. Summer months are the most popular. Daytime temperatures can be relatively high but usually there's a breeze off the lake. During winter months the breeze can turn into a gale and temperatures plummet but there are less people. However, any time you visit Fort Niagara is a good time.

www.oldfortNiagara.org/visitor-information

Major-General Friedrich Wilhelm Augustus, Baron von Steuben, a painting by Ralph Earl. (Courtesy of Yale University Art Gallery (CC0 1.0))

Friedrich Wilhelm August Heinrich Ferdinand von Steuben, who was born Friedrich Wilhelm Ludolf Gerhard Augustin Louis von Steuben in Magdeburg, Duchy of Magdeburg in 1730, had impressive bona fides. When he was 16 or 17, he enlisted in the Prussian Army and served as a second lieutenant during the Seven Years' War. He was wounded in 1757 at the battle of Prague and two years later was promoted to first lieutenant. That same year he was wounded for a second time in the battle of Kunersdorf and was named Deputy Quartermaster at general headquarters.

After he was taken prisoner at Treptow by the Russians, and exchanged, he became Major-General Von Knobloch's adjutant and later served as an aide-de-camp to Frederick the Great. At the end of the war, the army's ranks were greatly reduced and like many European military men Steuben found himself without a job, so he decamped for America reaching Portsmouth, New Hampshire, on December 1, 1777. He quickly became a celebrity and was fêted along with his aide-de-camp, Louis de Pontière, his secretary, Peter Steven Du Ponceau, and two other confidants in Boston.

Steuben's reputation had clearly preceded him. General Washington appointed him Inspector-General. Steuben went to work immediately, founding standards for camp layouts and hygiene. Equally importantly, he initiated a structure of progressive training and put sergeants in charge. This had never been done before. As part of that training he also emphasized the use of bayonet charges, and the battle of Stony Point was won largely because of a bayonet charge with unloaded muskets.

In the winter of 1778/9 Steuben wrote *Regulations for the Order and Discipline of the Troops of the United States* whose precepts were used by the U.S. Army until 1814 and had a positive effect on tactics and drills until the Mexican–American War of 1846. Steuben also sat on the panel that sentenced John André to death. He became an American citizen by an act of the Pennsylvania legislature in March of 1784 and two years later by New York authorities. He subsequently moved to upstate New York and lived on a small estate in Oneida County, where he died on November 28, 1794.

Many cities in the United States have Steuben Day parades, including New York and Chicago (featured in the film *Ferris Bueller's Day Off*). During World War I, the SS *Kronprinz Wilhelm*, a captured German ship, was renamed the USS *Von Steuben*, and in World War II a German luxury passenger liner was transformed into a transport ship, while during the Cold War a U.S. Navy submarine was named the USS *Von Steuben*.

| Fort Oswego

Fort Oswego is one of those forts whose name has to a great extent been subsumed by other forts in its vicinity. It was not built by an American but it nevertheless, in conjunction with its sister forts, played a small but important role in the Revolutionary War. The original fort was erected around a trading post on the lower ground on the northwest side of the mouth of the Oswego River. In 1727 a stone blockhouse was added to the existing structure and was named Fort Burnet. In 1741 a triangular stone wall 10 feet high (3 meters) and 3 feet thick (1 meter) was added and the enlarged structure changed names again, this time to Fort Pepperrell.

Fort Oswego plan. (National Archives and Records Administration (CC0 1.0))

Montcalm takes Fort Oswego, August 14, 1756, an illustration by John Henry Walker. (Bibliothèque et Archives nationales du Québec (CC0 1.0) {{PD-US-expired}})

In many ways, Fort Oswego refers in a general sense to the other forts in the immediate area that existed either at the same time or later. These forts often provided what would be known today as mutual aid. Fort Ontario began as a palisade on the high ground on the northeastern side of the river and Fort George was added to the bluff half a mile southwest from Fort Oswego. Fort George was also known as Fort Rascal or the West Fort, while Fort Ontario was known as the Fort of the Six Nations or the East Fort. The French knew Fort Oswego as Fort Chouaguen. This cluster of forts no doubt gave their respective garrisons a sense of security knowing help was not far away in case of attack.

Except this was not the case from August 10 through August 14, 1756, when Fort Oswego found itself alone, isolated, and under siege from a superior force. Even though the French had the only large naval force on Lake Ontario and could move freely between Fort Niagara and Fort Frontenac at the head of the St. Lawrence River, on March 27, 1756, they attacked Fort Bull on Wood Creek. Because the fort was an important depot in the chain that kept other forts supplied, they managed to destroy provisions intended for the Oswego garrison of 370 officers and men. Fort George had even fewer men: 150 New Jersey militiamen. Both forts were severely lacking in cannons.

A confident General Louis-Joseph de Montcalm led the French. He and the Governor of New France, Pierre de Rigaud, marquis de Vandreuil-Cavagnal, bickered over which fort to attack first, as French and Indian raiding parties under the command of Louis Coulon de Villiers started harassing the garrison at Fort Oswego.

The talented French general, Louis-Joseph de Montcalm, by Antoine Louis François Sergent dit Sergent-Marceau. (Library of Congress Geography & Map Division (CC0 1.0) {{PD-US-expired}})

A plan of Fort Ontario built at Oswego in 1759. (Library of Congress Geography & Map Division (CC0 1.0))

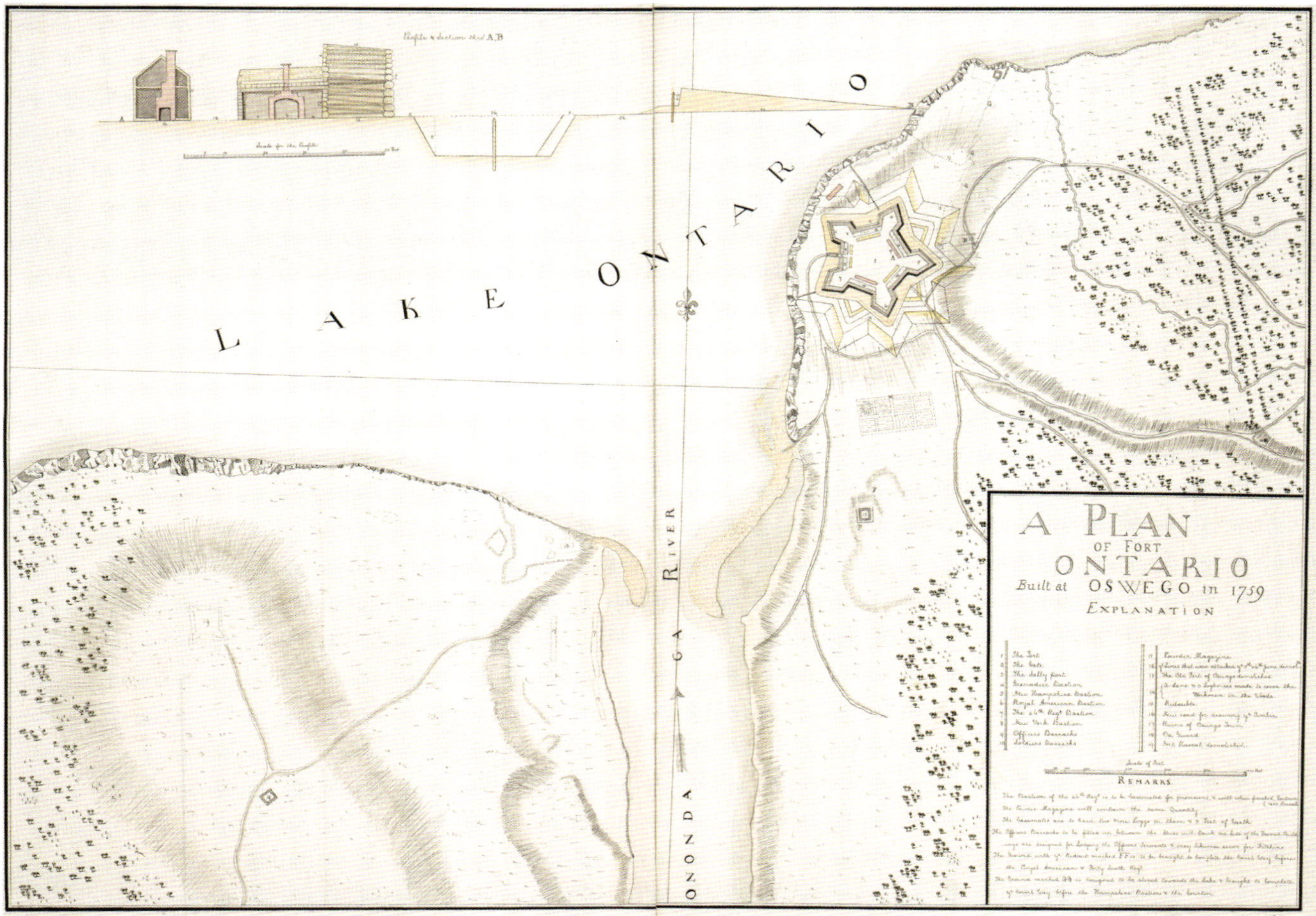

Forts Ontario and Oswego today, an aerial view. (Jacknayr (CC BY-SA 4.0))

French troops had gathered at Fort Frontenac and included the La Sarre, Guyenne, and Bearne battalions, marines, colonial militia, and approximately 250 Indians. The total force was around 3,000 men. On August 9 troops led by Rigaud and Villiers slogged overland toward Fort Oswego, while Montcalm and the remaining force disembarked approximately two miles (3.2 km) east of Fort Oswego.

Fort Oswego's fate was for all intents and purposes sealed. There would be no relief, no mutual aid, and it was only a matter of time before the fort fell to the French. Although the French were spotted by a British patrol boat who immediately called for reinforcements in the form of larger boats, these vessels were immediately driven off by French field artillery. Pierre Pouchot was tasked by Montcalm to ascertain how to best besiege Fort Oswego.

The initial gambit came at night on August 11/12 when the French opened siege trenches and started to laboriously work their way toward Fort Ontario. This was a well-established, proven tactic imbedded in French military doctrine. There was an exchange of rifle and cannon fire between the two sides until almost dusk of August 13. Then the commander of the fort, Captain James Mercer, a Virginian lawyer and later planter, jurist, and politician gave the order to abandon the fort, even though the hard-working French sappers had not as yet reached their goal. The experienced Montcalm knew exactly what he wanted to do and wasted no time in getting on with it. He immediately occupied the fort and built batteries on the western edge of the heights so his cannons could reach Fort Oswego's exposed east side. By the morning of August14 the French had nine guns in place that immediately opened fire on Fort Oswego's exposed and vulnerable stonework. The barrage reduced the stonework to rubble.

The wily Montcalm ordered Rigaud to take a number of men upstream, where they made an unopposed crossing and appeared at a clearing outside Fort Oswego. Almost simultaneously, the fort's commander, Colonel Mercer, was killed and Lieutenant-Colonel John Littlehales, who had taken over from him, raised a white flag in surrender on August 15. The entire operation cost the French 30 men either dead or wounded, while the British lost between 80 to 150 dead or wounded, plus 1,700 people including noncombatants taken prisoner. The fort's 21 cannons were seized as well. After the battle, the French destroyed the fort.

The fort today: Fort Oswego no longer exists and there is no website for it. The exact location is East 7th Street & Lake Ontario, Oswego, NY.

Visitors can go to Fort Ontario, 1 East Fourth Street, Oswego, NY. Amenities: Demonstrations, picnic area, reenactments, tours, and a visitor center where there are drawings and other images of Fort Oswego. The tours (call for prices) are informative and educational for all ages. The reenactments are exciting, as are the demonstrations such as cannon and musket firings.

This is a primer on the forts that comprise the greater Fort Ontario complex. They will add context to the events previously described. It sits on 36 acres and was added to the National Register of Historic Places in 1970. The current fortifications were built on a pentagonal plan not unlike that of Fort Jackson, Louisiana. It was designed for heavy cannon mounted *en barbette*, meaning they were situated over the parapet. Howitzers, however, were mounted in casemates built into the ramparts of the bastions. The stone-faced scarp walls of the bastions also had loopholes so that muskets could be fired through them. Initially, a ravelin protected the side of Fort Ontario that faced Lake Ontario. Visitors can see officer quarters, the powder magazine, the enlisted men's barracks, the storehouse and officer quarters 2. There are also two guardhouses by the entrance to the tunnel at the main entrance.

www.parks.ny.gov/historic-sites-fortontario/details.aspx

Old Stone Fort

The original structure was a Reformed Dutch Church built in 1772, in Schoharie, New York, west of Albany. At the onset of the Revolutionary War, a log stockade around the church was prudently added. The fort sits on 0.4 acres and was designed by John Schuyler. The word *schoharie* is derived from the Mohawk word for driftwood. The area had long been occupied by indigenous people, specifically the Mohawks, who were one of the Five Nations of the Haudenosaunee or Iroquois Confederacy; at its peak Iroquois power ranged as far north as the St. Lawrence River and as far east to the Hudson River.

The Old Stone Fort today. (Adam Lenhardt (CC BY-SA 3.0))

An 18th-century barn, Schoharie. (яісκy sнояе courtesy of Flickr (CC BY 2.0))

In 1713 Palatine Germans were the first Europeans to settle in the area after it was explored in 1710 through 1711. In 1710 approximately 3,000 German Protestant refugees sailed to New York, courtesy of Queen Anne's government. They were refugees from the religious warfare on the border with France, exacerbated by the loss of their crops in 1709 when the Rhine froze over. Believing the Germans could help in the development of the colony, the British granted them land west of English settlements.

On October 17, 1780, Sir John Johnson and the Mohawk firebrand Captain Joseph Brant led a force of around 800 Loyalists and Indians on a raiding party throughout the valley where the fort lay. They attacked the fort before they moved off in a northerly direction toward the Mohawk Valley. A cannonball hole can still be seen in a cornice at the rear of the building.

In 1785 the stockade was taken down and the building continued to function as a church until 1844 when the present-day Reformed Church replaced it. In 1857 it was sold to the state of New York for less than $1,000. Through the Civil War and until 1873 it was utilized as an armory. After that it was given to the county for historical use.

Although it did not have a garrison of musket-armed troops nor was bristling with cannons, the Old Stone Fort and others like it nevertheless played an important role as world powers sought to exert their influence and lay claim to unsettled territory. Sprinkled throughout the east coast and often subject to attacks either by French, British, or marauding Indians, these relatively tiny forts stood firm.

The Lutheran Parsonage, Schoharie, c. 1930. (Library of Congress Prints and Photographs Division (CC0 1.0))

The fort today: In 1888 the Schoharie County Historical Society was formed to operate a museum at the site of the old fort and by the following year a catalogue of 2,500 items was published. It was listed on the National Register of Historic Places in 2002. The Society has a collection of more than 50,000 artifacts relating to both the county and the fort. There is also a local genealogy collection and it publishes a biannual local history magazine. The Old Stone Fort is located at 145 Fort Road, Schoharie, NY 12157. The fort and other buildings sit on a 25-acre site that includes the William Badgely Museum and carriage house that was built in 1972, the Warner House, a Greek revival home that houses the Scribner Exhibit of 20th-century communications, the 1830 Jackson law office, the Oliver one-room schoolhouse that was furnished around 1900 and the Schaeffer-Ingold Dutch barn. Guided tours are available but are limited to eight people. Admission fees vary. The library is staffed by volunteers and is open subject to their availability. It is best to call before you go.

https://theoldstonefort.org.

Fort Salonga

Nothing remains of Fort Salonga, sometimes known as Fort Slongo after one of the fort's British contracted architects, George Slongo, who resided in Philadelphia, Pennsylvania. It was located near the border of Huntington township and the town of Smithtown that overlooks Long Island Sound. The fort was built in either 1778 or 1779 when British fort construction on Long Island was at its peak. The fort's location overlooked Long Island Sound and it was the smaller sister fort to the larger Fort Franklin. Salonga was armed modestly but sufficiently for its mission with two 4-pounder cannons, two 1-pounder cannons, and an unusual 1-pounder cannon made of brass, which was not the preferred metal used in making cannons because it is relatively soft.

Not all battles involve thousands of men firing muskets and cannons at each other. Many fights are much smaller affairs in which the adversaries number only in the dozens or hundreds, which is not to say they are less deadly or less bloody. On the contrary, because they are more intimate, every man who participates is much more likely to be in the thick of the action. Such a fight took place on October 3, 1781.

Historic marker for the battle of Fort Slongo, along eastbound New York State Route 25A on the Smithtown side of Fort Salonga, New York. (DanTD (CC BY-SA 4.0)

American Continental Army forces under the joint command of Benjamin Tallmadge and Lemuel Trescott engaged the British defenders at Fort Slongo. The Americans numbered 100 infantrymen, while the British garrison numbered between 80 and 140 infantrymen. Sergeant Churchill led an American reconnaissance force prior to the attack on Fort Slongo.

Lemuel Trescott was commissioned as a captain in Colonel John Brewer's Massachusetts Regiment on May 19, 1775. The next year he was a captain in the 6th Continental Regiment, also in Massachusetts, and in 1777 served in Colonel David Henley's Additional Continental Regiment. He was promoted to major on May 20, 1778, and, on April 22, 1779, was transferred to Colonel Henry Jackson's Additional. Two years later he was major of the 7th Massachusetts, and nominally on the Additional Continental Regiment roll, which on July 23, 1780, was titled the 16th Massachusetts. On January 1, 1781, he was transferred again, this time to the 9th Massachusetts, and led the raid on Treadwell's Mill on October 10, 1781.

Elijah Churchill enlisted in the 8th Connecticut Regiment as a private on July 7, 1775. Two years later, on May 7, 1777, he reenlisted for the duration of the Revolutionary War as a corporal in the 2nd Continental Light Dragoons, which was later named the 2nd Legionary Corps; he was promoted to sergeant on October 2, 1780. He was subsequently cited for gallantry in action in a fight at Fort St. George near Brookhaven, New York, on Long Island, in November 1780, and at Tarrytown, New York, in July 1781.

The modern Purple Heart. (USAF)

The Badge of Military Merit, the original Purple Heart. (U.S. Department of Defense)

As fate would have it, many of the British officers at the fort had been at a party the night before the battle. The commanding officer of the fort, Major Valentine, was away in New York City on official duty. Sergeant Churchill and his men beached their boats at Crab Meadow west of the fort and stealthily made their way to the nearby Nathanial Skidmore farm, where Skidmore led the reconnaissance party to the fort so they could plan the attack.

The fort today:

The fort no longer exists. Today it is an archeological site.

Led by Tallmadge and Trescott, on the night of October 2, 100 men climbed into their whaleboats at Norwalk, Connecticut, across the Long Island Sound. Truscott's force then split into two groups of 50 men each. Captain Richard's company of the Connecticut Line led one group and Captain Edger's Dismounted Dragoons the other. In the early morning hours of October 3 Edgar's dragoons were ordered to launch a surprise attack on the fort, while Captain Richard's infantry was ordered to surround the fort in order to prevent any members of the garrison from escaping.

The attack was launched at 3 a.m. It is reasonable to assume that the officers who had attended the party had overindulged and were sound asleep, as were the other members of the garrison (except for those troops on guard duty). Lieutenant Rogers of the 2nd Regiment of Light Dragoons led the attack; Major Trescott and Captain Edgar's men were behind him.

A British sentry spotted the approaching enemy and fired a shot that alerted the garrison. Then he scampered back inside the safety of the fort but in his haste or panic forgot to shut the gate behind him. Once the Americans gained entrance to the fort, resistance was slight and obviously futile. Casualties were light on both sides. The Americans had one man wounded while the British had four men killed, two wounded, and 21 captured. The Americans then burned all the buildings to the ground in addition to the stored materials, including the cannons. The fort was rendered unusable for the remainder of the Revolutionary War

The only American wounded in the attack was Sergeant Elijah Churchill, who would be awarded the Badge of Military Merit by George Washington. The badge became known as the Purple Heart that is still given to American soldiers who are wounded in action. Churchill was the first person to receive this medal. There is no British Army equivalent for the Purple Heart.

Fort Stanwix

Fort Stanwix was constructed under the direction of British General John Stanwix for whom the fort is named. It took four years from August 26, 1758, to 1762 to complete. It was a bastion fort, the mission or purpose of which was to guard a portage known during the French and Indian War as the Oneida Carry. This was between the main waterway that ran in a southeasterly direction to the Atlantic coast down the Mohawk River and the Hudson River, plus an interior waterway that ran in the opposite, northwesterly direction to Lake Ontario, down Wood Creek and Oneida Lake to Oswego.

In 1768 the fort was involved in the Treaty of Fort Stanwix, which established a line of property following the Ohio River that gave the Kentucky part of the Virginia colony and much of what is now West Virginia to the British Crown. The treaty also settled land claims between the Iroquois and the Penn family. The lands obtained by American colonists in Pennsylvania were known as the New Purchase. The treaty was negotiated between Sir William Johnson, his deputy George Croghan, and delegates of the Iroquois. The British hoped a new boundary line would end frontier violence, while the Indians hoped that a long-lasting line would deter white colonial expansion. Although the colonists got Kentucky, the tribes actually from Kentucky, such as the Shawnee, Delaware, and Cherokee, did not have representatives at the negotiations. Instead of securing peace, the treaty had the opposite effect.

This mammoth bastion fort was rectangular in shape and occupied 16 acres where the city of Rome, New York, is now located. The walls, constructed of 2-foot-squared logs, were an imposing 17 feet high (5 meters). A 40-feet-wide (21 meters) by 14-feet-deep (4 meters) dry moat surrounded the fort. The bastions had a flank length of 35 feet (10.6 meters) and a face length of 108 feet (32.9 meters). Bombproof, 16-feet-square, dirt-floored rooms in each bastion were protected by a single layer of heavy roof planks and were 3 feet below the parade ground that averaged 451 feet (137 meters) above mean sea level. The southeastern bastion was 21 feet

Fort Stanwix, taken in 2014.
(M. Colangelo Sr. (CC BY-SA 3.0))

Fort Stanwix National Monument, Rome, NY. (NPS Photo www.nps.gov/common/ (CC0 1.0))

long (6.4 meters) on the eastern side, 20 feet long (6.1 meters) on the western side, 19 feet 6 inches long (6 meters) on the northern side, and 18 feet 5 inches long (5.6 meters) on the southern side. Two comparatively roomy barracks housed officers and enlisted soldiers. The eastern barracks was 20 feet wide (6 meters) by 110 feet long (33.5 meters), while the western barracks was 120 feet long (36.5 meters) by 20 feet wide (6 meters). The bake house could produce 200 loaves a day which meant that each member of the garrison received a full ration.

John Stanwix joined the army in 1706 and steadily rose through the ranks. In 1739 he was a captain in the Grenadiers; two years later he was promoted to major in the Marines, and four years after that he was a lieutenant-colonel and equerry to Frederick, Prince of Wales. In 1750 he was appointed Governor of Carlisle and was also a Member of Parliament representing Carlisle. He became Deputy Quartermaster in 1754 and on January 1, 1756, he was promoted to Colonel-Commandant of the 1st Battalion of the 60th Regiment that was also known as the Royal American Regiment. When he arrived in America he was given command of the southern district and during 1757 he was coincidentally headquartered at Carlisle, Pennsylvania, where, on December 27 of that year, he was promoted to brigadier-general. In 1758 he was sent to Albany, New York, to the Oneida Carrying Place to begin construction of the fort. He was made a lieutenant-general on January 19, 1761. When he returned to England, he was named Lieutenant-Governor of the Isle of Wight and was Colonel of the 49th Regiment of Foot from 1761–4 and the 8th (King's) Regiment of Foot from 1764–6. He was also a Member of Parliament for Appleby in Westmoreland from 1761–6. He was lost at sea while crossing from Dublin, Ireland, to Holyhead, Wales, while aboard the packet *The Eagle*.

Fort Stanwix. (Ernest Mettendorf)

Fort Stanwix fortifications overlooking the entrance bridge and moat. (National Park Service Digital Image Archives (CC0 1.0))

American troops (reenactors) in their quarters at Fort Stanwix (National Park Service Digital Image Archives (CC0 1.0))

Stanwix commanded a battalion of approximately 600 troops. However, that number fluctuated due to death, illness, casualties and the time it took for replacements to arrive. Fort Stanwix had an impressive array of cannon. The 3-pounder was known as the "Galloper" and was capable of launching a three-pound iron ball; it was used extensively as was the 6-pounder. These could fire diverse rounds such as solid shot, grapeshot, chain shot and explosive-filled balls. Maximum range was 1,000 yards (91 meters) and a capable, well-trained crew of 10 could fire four rounds a minute. Mortars with a shorter barrel and a higher firing angle were also used. Their maximum range was roughly 1,400 yards (1,280 meters) but they were only effectual up to 700 yards (640 meters). They were offensive as opposed to defensive weapons, as were howitzers whose calibers were larger than cannons. Their length was a stumpy 3 feet (1 meter) and they had an operational range of around 750 yards (685 meters).

In 1758 General Stanwix was replaced by Brigadier-General John Forbes, a professional Scottish soldier who had served in the British Army for 30 years. During the French and Indian War, he led the Forbes Expedition in 1758 that occupied the French outpost Fort Duquesne and constructed Forbes Road, a significant road for settlement of the Western United States. Prior to that, he fought in the War of Austrian Succession.

Fort Stanwix was vacated in 1768. Eight years later, under the command of Colonel Elias Dayton, the fort was occupied by colonial troops and reconstructed. Renamed Fort Schuyler, it was commonly called Fort Stanwix. But it was Colonel Peter Gansevoort who assumed command of Fort Stanwix/Schuyler on May 3, 1777, and who would be the object of a vicious three-week-long siege led by Barrimore Matthew "Barry" St. Leger. Gansevoort's fort totaled 700 men who received their final boatload of supplies on August 2 while taking fire from St. Leger's advance force.

General Peter Gansevoort, an oil painting by Gilbert Stuart. (Courtesy of Munson-Williams-Proctor Institute of Art (CC0 1.0) {{PD-US-expired}})

Peter Gansevoort was a more than capable commander. When the Revolutionary War began, Gansevoort joined the Continental Army and was made a major on June 30, 1775, and served as a field commander in the 2nd New York Regiment. He led the regiment during the siege of Fort St. Jean, Quebec. He captured Fort Chambly and over 120 barrels of gunpowder including a huge mortar they named the "Old Sow", plus about 100 prisoners of the Welch Fusilier garrison and their captain, John André. Gansevoort then participated in the capture of Montréal. In the spring of 1776 he stopped a British advance on Lake Champlain and in June, as a reward, was assigned command of Fort George. His area of responsibility was from the Hudson River valley and Fort Edward and Fort George along the Mohawk River valley to Fort Oswego in the northwest. He wisely conceded Fort Oswego to the British and decided to defend Fort Stanwix instead, with a garrison of 600 officers and men. When the war ended there was no assignment for him in the army. He became brigadier-general of the Albany County Militia. He also served as sheriff of Albany County and as a commissioner of Indian Affairs. In 1809 he was a brigadier-general in the United States Army and commanded the Army's Northern Department. He died of a recurring illness in the summer of 1812.

Sleeping quarters at Fort Stanwix. (M. Colangelo Sr. (CC BY-SA 3.0))

Colonel Barry St. Leger's three-pronged campaign was intended to divide the American colonies and put an end to the rebellion. St. Leger commanded 1,800 men who were a composite of British regular soldiers, Hessian Jägers, Rangers, Loyalists, and Mohawk and Seneca of the Iroquois Indians. Their route had taken them up the St. Lawrence River and along Lake Ontario's shore to finally reach the Oswego River. Then they climbed to the Oneida Carry in Rome, New York, where Fort Stanwix oversaw the portage.

Barry St. Leger had an impressive resumé, like the vast majority of British officers who saw duty during the Revolutionary War. In April 1756 he joined the British Army and served as an ensign in the 28th Regiment of Foot. In 1758 he was present at the siege of Louisbourg during the French and Indian War, and at the siege of Quebec. In July 1760 he was a brigade major in the campaign against Montréal. Two years later, on September 16, 1762, he was promoted to major in the 95th Regiment of Foot. At the onset of the Revolutionary War, he was a lieutenant-colonel in the 34th Regiment of Foot. He was a principal player at the battle of Oriskany.

Barry St. Leger. (REA https://archive.org/ (CC0 1.0))

On August 2 the siege by the King's 8th Regiment of Foot, Loyalists and Indians began in earnest when Gansevoort refused to formally surrender Fort Stanwix. In an act of brazen defiance Gansevoort raised the American flag over Fort Stanwix. It was the first time that the flag of the United States was flown in battle. St. Leger made good use of his cannons and howitzers and almost incessantly battered the solid, protective wooden walls of the fort. Musket fire cracked through the air from both sides. Defenders and attackers alike sought cover whenever and wherever they could. Acrid smoke filled the air and wafted skyward. The noise at times was deafening. Wounded men cried out in pain. Medical attention was scant.

Gansevoort stubbornly refused to cede control of the fort and rallied his men in spite of the almost continuous bombardment and gunfire. In spite of this the guns of Fort Stanwix continued to fire, their crews disregarding the withering fire they were under. The defenders drank, ate, and slept at their posts on the walls. Dead and dying attackers lay in the moat. There was little respite. It was clear to both the Americans and the British that the fort could not hold out indefinitely. The garrison's strength had been severely compromised and ammunition was bound to run out sooner or later. All St. Leger had to do was to keep pounding with his guns while skirmishers kept defenders' heads down and reduced their ability to return fire. It was just a matter of time before the fort fell.

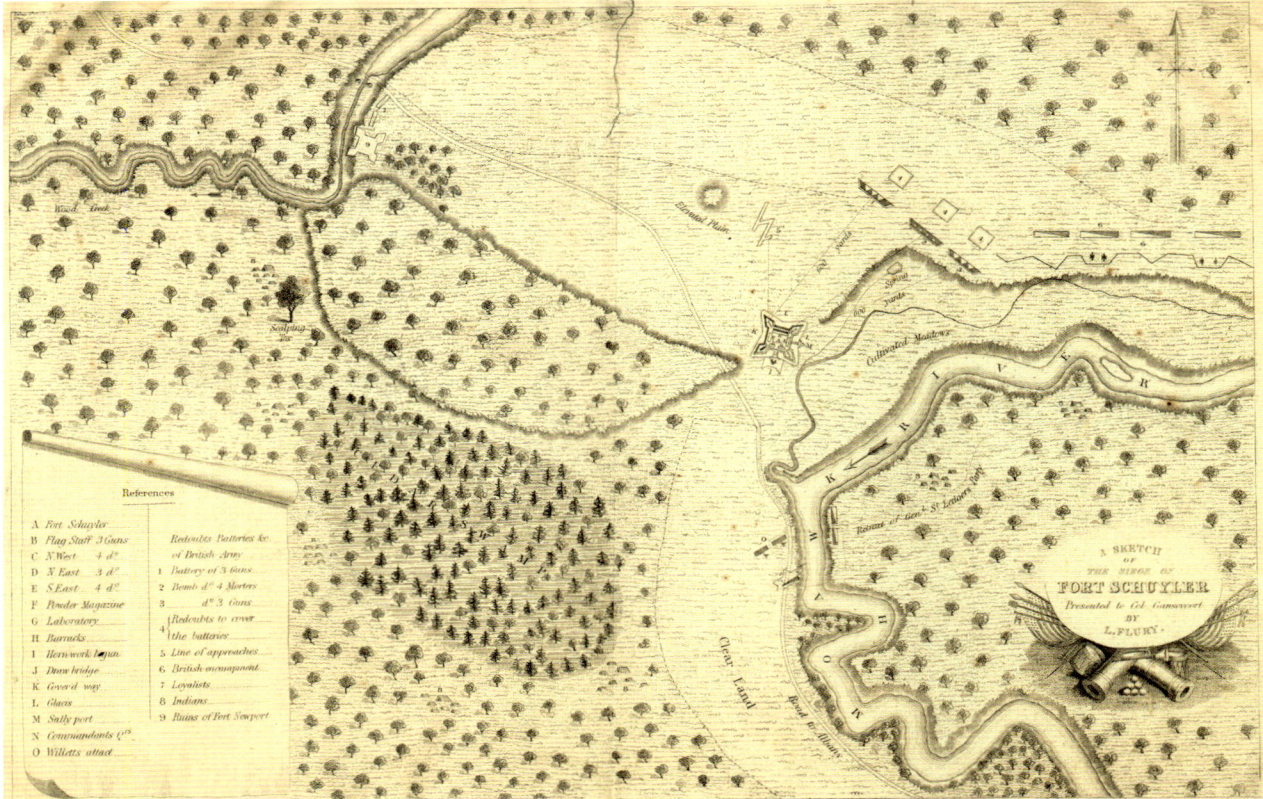

Sketch of the siege of Fort Schuyler, presented to Colonel Gansevoort by the artist Francois Louis Tesseidre de Fleury, 1777. This map depicts the positions and movements at the siege of Fort Stanwix (also known as Fort Schuyler). (Courtesy University of Pittsburgh Digital Collection (CC0 1.0))

Unbeknown to both St. Leger and Gansevoort, help was on the way in the form of a relief column. Upon learning that Fort Stanwix was being besieged, in late July, Brigadier-General Nicholas Herkimer ordered his 800-man strong Tryon County Militia, augmented by some 60 Oneida Indians, to assemble at Fort Dayton. Fort Stanwix lay around 28 miles (45 kilometers) to the west.

When he learned of this, St. Leger acted promptly, sending a force of 450 men including a light infantry (Hanau Jäger) detachment, Sir John Johnson's King's Royal Regiment of New York, Indian Department Rangers, and Mohawk and Seneca Indians to intercept the relief column before it could reach the fort. The battle of Oriskany would shortly begin and it would prove to be one of the bloodiest fights of the Revolutionary War.

Herkimer's column was ambushed about six miles (10 kilometers) east of the fort in a small valley near Oriskany, an Oneida village. It was the perfect place to set a trap. The track descended more than 50 feet into a swampy gulley about a yard wide. The plan was originated by the Seneca chiefs Sayenqueraghta and Cornplanter. The Indian contingent positioned themselves on either side of the gulley, concealing themselves in thick brush and behind trees while the King's Royal Yorkers were on a small rise. It was they who were tasked with stopping the head of Herkimer's column while the Indians opened fire on the rest of the drawn-out column.

Normally the head of the column would have been attacked first to stop it dead in its tracks and prevent it from moving forward. Instead, the Indians inexplicably fired on the rear of the column. After the commander of the 1st Regiment, Colonel Ebenezer Cox, was killed in the first volley, a startled Herkimer turned to see what was happening and was hit by a musket ball that not only killed his horse but also shattered one of his legs.

The siege of Fort Stanwix (reenactment). (M. Colangelo Sr. (CC BY-SA 3.0))

The men who hadn't entered the gully promptly fled, pursued by Indians who left a trail of dead and dying for miles. It was estimated that after 30 minutes, only 50 percent of Herkimer's troops were still fighting, the rest having been killed in the initial volley or were being pursued while fleeing. Many attackers were armed only with tomahawks that proved lethal, as it took time for Herkimer's men to reload their muskets.

Johnson raced to the British camp where he requested reinforcements and was given 70 men. A fortuitous downpour halted the fighting for about an hour, allowing Herkimer to regroup his remaining men on higher ground. He sent a runner to the fort requesting much needed relief. At 11 a.m. Gansevoort sent Lieutenant Marinus Willet and 250 soldiers from the fort so they could raid and pillage the almost completely deserted British camps that lay to the south.

When the Indians who were attacking the Americans found out that their camps were being looted and destroyed, many left the battle to try and save their women and belongings. German and Loyalist soldiers also withdrew from the battle when they saw their Indian comrades leave.

When he learned of the approach of another relief column, led by Benedict Arnold, and another column led by General Philip Schuyler, St. Leger prudently withdrew his forces through Canada so he could join General Burgoyne's campaign at Fort Ticonderoga.

General Philip Schuyler led one of the relief columns to Fort Stanwix. Painted by Jacob H. Lazarus from a miniature painted by John Trumbull. (Courtesy of Schuyler Mansion State Historic Site, Albany (CC0 1.0) {{PD-US-expired}})

A cannon night shoot at Fort Stanwix (reenactment). (Kent Bolke (CC BY-SA 4.0))

The battle at Oriskany almost annihilated Herkimer's garrison. Nearly half were either killed or wounded (385 killed, 50 wounded, and 30 captured). By way of contrast, only 15 percent of the British participants were killed or wounded. The Indians lost 65 men either killed or wounded, while the British lost seven killed and 21 wounded, missing, or captured. The American brigade surgeon, William Petrie, dressed Herkimer's wounds in the field, but the decision to amputate his leg was delayed for 10 days because Petrie himself had been wounded. The operation was performed by the less-experienced surgeon Robert Johnson who botched the operation. The 49-year-old Herkimer died on August 16.

The battle marked the onset of a war between the Iroquois as Oneida warriors led by Colonel Louis and Han Yerry allied themselves with the Americans, and the Mohawks and Senacas allied themselves with the British. The site of the battle is known in Iroquois oral histories as "A Place of Great Sadness."

The guns of Fort Stanwix and the crews that manned them were victorious, a testament to not only leadership but the durability of the fort's cannon that withstood use not normally anticipated. On May 13, 1781, the fort suffered an ignominious end. It was burned to the ground and not rebuilt. The garrison moved to Fort Herkimer. The second Treaty of Fort Stanwix was conducted at the fort between the Americans and the Indians in 1784. During the War of 1812, a blockhouse was built on the parade ground. In early 1828, the fortifications were dismantled.

The fort today:

Fort Stanwix was declared a National Monument on August 21, 1935. It was added to the National Register of Historic Places on October 15, 1966. It was designated a National Historic Landmark on November 23, 1962. The National Park Service reconstructed the fort between 1974 and 1978, and in 2005 a new visitor center was added. The fort sits on 16 acres and is located at 100 North James Street, Rome, NY. There are guided and self-guided tours but during the winter months there are only ranger-guided tours. Reenactors and the replication of the fort will give visitors of all ages instructional insights into the daily life of the garrison. For aficionados of Revolutionary War forts, this is a must-see place to visit.

Fort Ticonderoga

The French called it Fort Carillon. It occupied a strategic position near the southern end of Lake Champlain that controlled a portage next to the La Chute River. The portage was three and a half miles (5.6 kilometers) long, between Lake Champlain and Lake George, and thus at the nexus between trade routes of the British-controlled Hudson River valley and the French-owned St. Lawrence River valley. The name Ticonderoga derives from the Iroquois word *tekontaro:ken* that translates as "it is at the junction of two waterways."

A plan of the town and Fort of Carillon at Ticonderoga: with the attack made by the British army commanded by Genl. Abercrombie, 8 July 1758, by Thomas Jefferys. (Map reproduction courtesy of the Norman B. Leventhal Map & Education Center at the Boston Public Library, https://collections. leventhalmap.org/search/commonwealth:9s1618735)

A view of the lines and Fort Ticonderoga taken from a hill on the side of South Bay in 1759. (Library of Congress Prints and Photographs Division (CC0 1.0) {{PD-US-expired}})

The fort was designed by Michel Chartier de Lotbinière. Construction started in October of 1755 and went smoothly but slowly during the warmer months of 1756 and 1757. Troops from the nearby Fort St. Frédéric and those stationed in Canada were used. The fort was smaller than the typical Vauban-style fort. It was 500 feet (150 meters) wide. Its barracks could accommodate a garrison of 400 soldiers. It also had a small cistern. The French named it Fort Carillon because the churning rapids of La Chute River (known today as Ticonderoga Creek) sounded like the chiming bells of a carillon.

Michel Chartier de Lotbinière, marquis de Lotbinière, was born in Quebec in 1723. In 1746 and 1747 as a second ensign he served in the defense of the Acadians. The marquis had influential relatives. In 1747 he married Louise-Madeleine Chaussegros, the daughter of Gaspard-Joseph Chaussegros de Léry, the chief engineer of New France. Another relation, Roland-Michel de la Galissonière, Commandant-General of New France, promoted him to ensign so he could train him as an engineer and artillery officer. In 1753 he was a lieutenant with the title of King's Engineer in the Colonial Regular Army and worked with his father-in-law on the construction of the ramparts of Quebec City. In 1755, his cousin, Pierre de Rigaud, marquis de Vandreuil-Cavagnal, put him in charge of building Fort Carillon/Ticonderoga.

Lotbinière, from *Histoire des Canadiens-Français. 1608–1880. Ouvrage orné de portraits et de plans*, by Benjamin Sulte. (Courtesy of the British Library and Flickr (CC0 1.0))

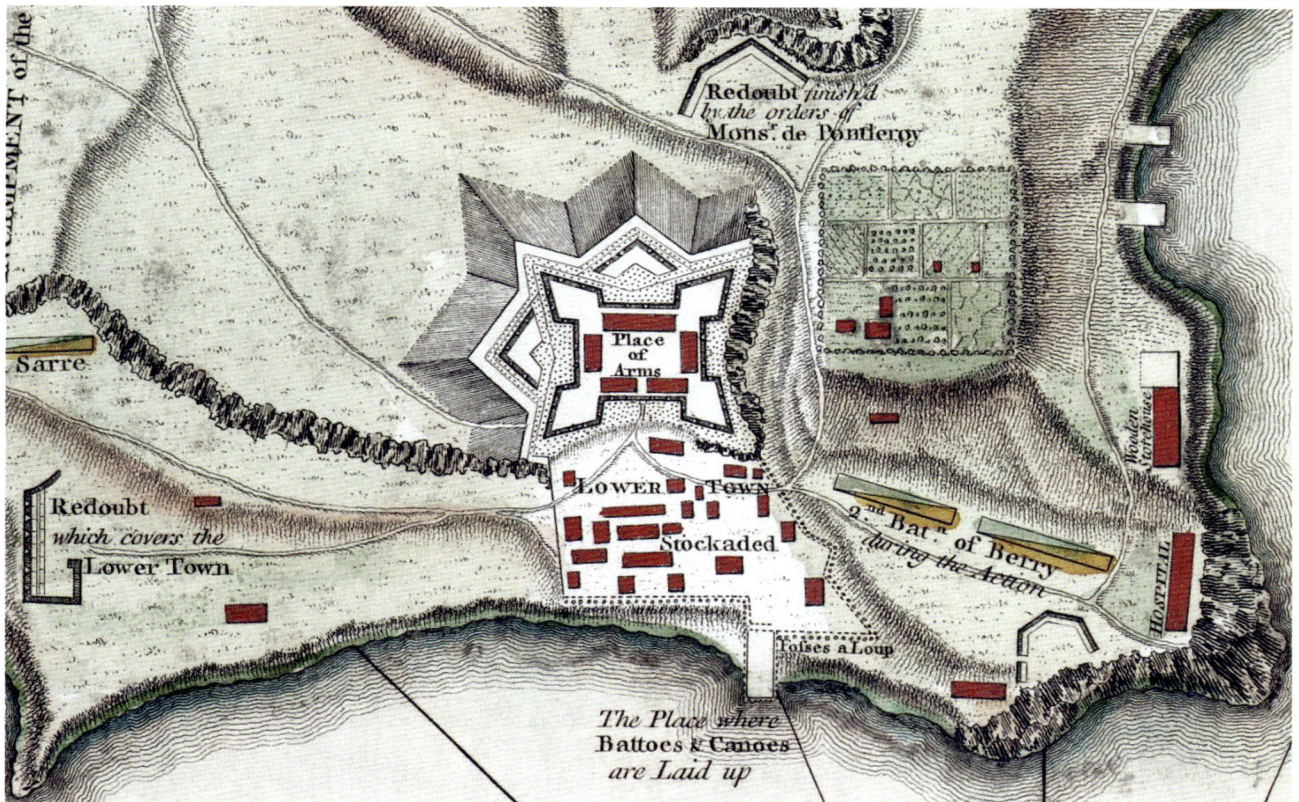

Fort Ticonderoga plan, 1758. (Thomas Jefferys Publisher, edited/restored by Matt Wade and modified by UpstateNYer (CC0 1.0))

The walls were 7 feet (2.1 meters) high and 14 feet (4.3 meters) thick and were made of squared wooden timbers with dirt filling the gaps. The fort was surrounded by a glacis and a dry moat 5 feet (1.5 meters) deep and 15 feet (4.6 meters) wide. The walls were also reinforced with stone from a quarry a mile away. The fort had three barracks and four storehouses. One of the bastions held a bakery that could produce 60 loaves of bread a day. Each loaf weighed 6 pounds (2.7 kilograms). Beneath another bastion a powder magazine was made by removing bedrock. Cannons were hauled from Montréal and Fort St. Frédéric. The area outside the fort was protected by a wooden palisade between the southern wall and the lakeshore. This space was the main landing area for the fort and contained more storage facilities needed for the maintenance of the fort, as storage space in the fort was limited. The fort had one critical strategic weakness: several nearby hills overlooked the fort, enabling attackers to fire down on it.

On January 21, 1757, occurred an interesting little skirmish aptly called the battle on Snowshoes. Captain Robert Rogers and his 74 Rangers were on a scouting expedition near the fort when they were ambushed. Rogers had seen a supply convoy of sleds being lugged toward Fort Carillon and gave chase. The leader of the supply convoy made it safely to the fort and reported to its commander, Paul-Louis de Lusignan, who immediately dispatched a force of 90 regulars plus 90 Canadian militia and Ottawa Indians under the command of Captain de Basserode. The Indians were led by Charles Michel de Langlade. The ambush was set. However, because their gunpowder was wet, many of the French muskets misfired. The fight lasted several hours and only ended when the opposing sides could no longer see each other due to darkness and heavy snowfall. The British had a slight advantage because they wore snowshoes. Casualties were proportionally high. The French lost 11 men killed and 27 wounded, while the British lost 14 men killed, nine wounded, and six either captured or missing in action. A much bigger and bloodier battle was to follow 18 months later.

Fort Ticonderoga from Mount Defiance, taken in 2009. (Mwanner (CC BY-SA 3.0))

The French wasted no time in making good use of their new fort. In August 1757 troops from the fort captured the British Fort William Henry; this prompted the British to retaliate with a large-scale attack on Fort Carillon the following summer as part of a wider campaign against French Canada. The antagonists were the French, commanded by Louis-Joseph de Montcalm and the Chevalier de Lévis who fielded a force of 3,600 regular troops, militiamen, and Indians, and the British, commanded by James Abercrombie, with a force of 6,000 regular troops plus 12,000 provincial troops, Rangers, and Indians.

The portage trail could be traced from the northern end of Lake George to the location of a sawmill the French had built for the fort's construction. It crossed the La Chute River twice by means of two bridges, one about two miles (3.2 km) from Lake George and again at the sawmill that was the same distance from the fort. Mount Defiance, known at the time as Rattlesnake Hill, with an elevation of 900 feet (270 meters), was south of the fort on the other side of La Chute River. The sides of the hill were steep and densely wooded, excellent locations to position cannon to fire directly down onto the fort.

British troops disembarked from boats at the northern end of Lake George on July 6, 1758, and included Lord John Murray's Highlanders of the 42nd Regiment of Foot, the 27th (Inniskilling) Regiment of Foot, the 44th Regiment of Foot, the 46th Regiment of Foot, the 55th Regiment of Foot, the 1st and 4th Battalions of 60th (Royal American) Regiment, and Colonel Thomas Gage's Light Infantry. Provincial militia was from Connecticut, Massachusetts, New York, New Jersey, and Rhode Island.

Because he had only 3,600 troops at his disposal who in turn had rations for nine days, Montcalm decided to defend the fort from its obvious approaches. His first move was to dispatch three battalions under the fort commander Colonel François-Charles Bourgamaque to fortify the river crossing on the portage trail about 2 miles (3.2 km) from the northern end of Lake George and about 6 miles from the fort itself. Montcalm took two battalions and fortified an advance camp at the sawmill. On July 5 Captain Trepezet and 300 men deployed

Plan of Fort William Henry on Lake George, by W. Eyre. Eng'r, I. Heath. D'r. (Courtesy of Library of Congress, Geography and Map Division. (CC0 1.0) {{PD-US-expired}})

General Sir James Abercrombie, a portrait by Alan Ramsey c. 1759. It all went horribly wrong for Abercrombie at Ticonderoga. (Courtesy of Fort Ligonier (CC0 1.0) {{PD-US-expired}})

to Lake George to conduct a reconnaissance of a British fleet that was approaching the fort. If they could, they were instructed to halt their advance. Bourgamaque was ordered to retreat and both bridges on the portage trail were destroyed, isolating Captain Trepezet and his men from the main French force.

On the evening of July 6 entrenchments on the rise three-quarters of a mile northwest of the fort were established. Below these entrenchments another line of defense in the form of an abatis was constructed, while above the entrenchments was a wooden breastwork. These hastily constructed defenses would have given the defenders protection from muskets but not from cannon (that the British ultimately chose not to use).

The British force landed unopposed and proceeded to march in columns up the western side of the stream that connected Lake George to Lake Champlain as opposed to using the portage trail, the bridges of which Montcalm had wisely destroyed. Unfortunately for the British, the woods on the western side of the stream were too thick to maintain the integrity of the columns.

Fort Ticonderoga cannon, taken in 2017. (Manuela Michailescu (CC BY-SA 4.0)

Thomas Gage, see here as a general, a portrait by John Singleton Copley, 1788. (Yale Center for British Art, Paul Mellon Collection (CC0 1.0))

Captain Trepezet and his men were still trying to reach the safety of the fort; where Bernetz Brook enters the La Chute River, he inadvertently ran into a Connecticut regiment led by Phineas Lyman, and a skirmish immediately ensued. General Howe's column was near the action and when they heard the musket fire they advanced in the direction of the firefight. As they got close, General Howe was hit by a musket ball and instantly killed. A column of Massachusetts provincials then proceeded to encircle Captain Trepezet and his men, killing half outright and capturing the rest. Trepezet escaped by swimming across the La Chute River but died the next day from the wounds he had suffered in the fight.

On July 7, 1758, Lieutenant-Colonel John Bradstreet led a large number of men down the portage path where they rebuilt the first bridge. The main army followed and set up camp at the sawmill. Abercrombie's engineer, Lieutenant Matthew Clark, was sent to scout the French defenses. Clark and his men ascended Mount Defiance that was 853 feet high (260 meters), while Mount Hope and Mount Independence also overlooked the French defenses. He then reported that the French defenses appeared to be incomplete, which they were not. Instead, the French had camouflaged them. On the basis of that report, Abercrombie decided to attack the next day. Just as dawn was breaking Clark again returned to Mount Defiance to reaffirm his initial assessment. He saw nothing that made him change his mind. He was confident that cannons were not needed.

On the morning of July 8 Rogers' Rangers and elements of Gage's 80th Regiment of Light Armed Foot successfully forced the French scouts behind the entrenchments. Three columns of regular British troops followed. The right-hand column was led by Irish-born Lieutenant-Colonel William Haviland, who commanded the 27th and 60th Regiments of Foot. The 44th and 55th, under the command of Lieutenant-Colonel John Donaldson, made up the center column, while Lieutenant-Colonel Francis Grant and the 42nd and 46th were on the left. Each column was preceded by the regimental light infantry companies. Provincial regiments from Connecticut and New Jersey were held in reserve.

Montcalm had organized the French defense into three brigades and a reserve. He commanded the Royal Roussillon and Berry Battalions that were positioned in the center of the entrenchments. Lévis commanded the Bearn, Guyenne, and la Reine Battalions on the right, while Bourgamaque and the La Sarre and Languedoc Battalions were on the left. Each battalion was tasked with defending 100 yards (91 meters) of entrenchment with the flanks protected by cannons, aside from the right flank, the guns of which were not in place. Militia and marines guarded the low ground and they too had built abatis. Reserve troops were either inside the fort or between the entrenchments and Mount Hope. Portions of each battalion were held in reserve, ready to reinforce French positions if and when they were needed.

British muskets were loaded. Sergeants kept the troops marching at a steady pace, barking encouragement and curses in equal measure. The numerical superiority of five to one gave Abercrombie a sense of invincibility that he would prevail even without cannon support. Fort Carillon was his for the taking.

Except that it wasn't.

Abercrombie believed the battle was going to start at around 1 p.m.. However, at 12:30, troops on the left from the New York regiments began engaging the enemy, causing Abercrombie to think that the French line had been penetrated if not broken. Even though all his well-trained regular troops weren't in place, he

The view of Fort Ticonderoga from the summit of Mount Defiance. (Mwanner (CC BY-SA 3.0))

The battle of Carillon, by John
Henry Walker, 1877. (www.
patrimoine-culturel.gouv.qc.ca
(CC0 1.0))

John Bradstreet, seen here as a
general, a painting by Thomas
McIlworth. (Courtesy of the
National Portrait Gallery (CC0
1.0) {{PD-US-expired}})

nevertheless ordered them to advance. The right went first followed by the center, and then the left. Instead of a well-organized advance, troops entered the battle in piecemeal fashion. The 42nd then entered the fight.

The abatis did their intended job and ground the British advance almost to a halt. Men in the front ranks struggled to penetrate the sharpened branches. Belts and straps got entangled. The French then unleashed withering enfilading fire on the British, turning the abatis into a killing field. Men screamed as they fell. Officers and sergeants tried to rally their men but to no avail. The French fire was too intense. Seeing this, Montcalm took off his coat and moved freely among his men, encouraging them as he realized the British attack was failing.

By 2 p.m. it was obvious the attack had failed. Dead and dying British soldiers littered the abatis and entrenchments. Inexplicably, Abercrombie persisted with further attacks and ordered the New Jersey and Connecticut reserves into action, but their efforts also failed.

On the La Chute River at about the same time the doomed attacks were underway, several British barges carrying much-needed cannons floated down a channel between a small island in the river and the shore. Unfortunately, this brought them well into range of the French left and Fort Carillon's cannons, two of which on the southwestern bastion promptly sank two of the barges, causing the others to beat a hasty retreat.

It took Abercrombie around 30 minutes to realize that the additional assaults had failed; he tried to correct his mistake by recalling the troops but a large number of soldiers from the 42nd (Black Watch) and 46th Regiments on the British left continued doggedly to attack. At approximately 5 p.m. the remaining members of the 42nd made one last, desperate assault and managed to actually reach the base of the fort's wall. The soldiers who somehow scaled the breastwork were bayoneted by French defenders. The slaughter continued until darkness mercifully fell. By this time, a large number of British troops had retreated and sought safety behind a breastwork at the rear of the battlefield. Abercrombie then ordered what was left of his force to march through thick woods to the landing on Lake George. Dawn saw the British rowing boats and by nightfall they had reached their base at the southern end of the lake.

For his part, Montcalm had sustenance brought to the men in the lines outside the fort. Anticipating yet another attack, they worked through the night on the defenses, including building two redoubts to the northeast of the fort, named Germain and Pontleroy for their engineers.

Abercrombie never led another military campaign and was regarded as a totally incompetent military commander. He was recalled to England several months later and in 1772 was promoted to full general. The balance sheet for the battle of Carillon made for unhappy reading: the French lost 100 men killed, 500 wounded, and 150 taken prisoner. British losses, that Abercrombie allegedly underreported, were horrific: 1,000 killed, 1,500 wounded, and 150 missing. More than 300 men of the Black Watch, including nine officers, were killed, and an equal number were wounded.

General Montcalm rallies his troops in the defense of Fort Carillon. (Canadian Military Heritage, Department of Defence (CC0 1.0))

Montcalm's troops celebrate victory, by Henry Alexander Ogden. (CC0 1.0)

Because the British did not attack again in 1758, the French reduced the fort's garrison to only 400 men and officers. In 1759 the fort was captured by 11,000 British troops led by General Jeffery Amherst, First Baron Amherst, who used emplaced cannons. The French gutted the fort of anything valuable and although the British worked for the next two years to repair and improve the fort, it never again played a significant role.

In the summer of 1776 Mount Independence, which is almost completely surrounded by water, was fortified with trenches, a horseshoe-shaped battery halfway up the hill, a citadel at the summit, and redoubts armed with cannons on the summit. A pontoon bridge guarded by land batteries on each side linked these defenses to Fort Ticonderoga. General Horatio Gates supervised the operation.

In March 1777 General Philip Schuyler, who had worked on the defenses before Gates, requested 10,000 troops to guard the fort and an additional 2,000 to protect the Mohawk River valley from a possible British invasion from the north. Washington disagreed and nothing was done to add to the fort's defenses. The garrison numbered 2,000 men and was commanded by General Arthur St. Clair.

In June 1777, 7,800 British and Hessian troops from Quebec were led south by General Burgoyne. After he occupied Fort Crown Point, he made plans to lay siege to Fort Ticonderoga and hauled several large-caliber cannons to the top of Mount Defiance. Aware that he was at a tactical disadvantage and outgunned and outmanned, General St. Clair abandoned the fort on July 5, 1777. It was then garrisoned by 700 British and Hessian troops under Brigadier-General Henry Watson Powell. Most of the troops were positioned on Mount Independence, while 100 were stationed at the fort and a blockhouse that was under construction on top of Mount Defiance.

After Burgoyne was defeated in October 1777 at the battle of Saratoga, Fort Ticonderoga became redundant and was abandoned by the British along with Fort Crown Point in November. British raiding parties occasionally used the fort, but its strategic value no longer existed. In 1781, after the British surrendered at Yorktown, they left the fort and never came back. Local residents stripped the fort of anything usable, including some of the cannons which were melted down and repurposed.

The storming of Fort Ticonderoga, 1775, an illustration by Frederic Remington. (New York Public Library and Flickr (CC0 1.0))

General Horatio Gates, from an oil painting by Gilbert Stuart. (Courtesy of Metropolitan Museum of Art (CC0 1.0) {{PD-US-expired}})

The fort today: Fort Ticonderoga is located at 102 Fort Ti Road, Ticonderoga, NY 12883. The fort was designated a National Historic Landmark on October 9, 1960, and added to the National Register of Historic Places on October 15, 1966. The fort sits on 2,000 acres and is compliant with the Americans with Disabilities Act. Pets are allowed but must be leashed. There is an extensive schedule of events. Wear comfortable shoes and remember that summer months can be quite warm and winter months are usually very cold.

www.fortticonderoga.org.

Ruins of Ticonderoga. (Library of Congress)

Fort Wadsworth

Fort Wadsworth was strategically situated on Staten Island on the Narrows that divide New York Bay into upper and lower halves. It was an obvious point from which to defend the upper half and Manhattan. Like other forts, it has undergone several name changes over the years. Fort Wadsworth is essentially the umbrella name for a collection of forts, all of which are in close proximity to one another. Although they are separate entities, these forts are collectively known as Fort Wadsworth.

The earliest use of this land for military purposes was in 1663 when a Dutch settler named David Pieterszen de Vries built a blockhouse on Signal Hill, now the site of Fort Tompkins. During the Revolutionary War the area was known as Flagstaff Fort that was captured by the British in 1776, though they relinquished it in 1783. In 1806 New York State took responsibility for the area and initiated the building of four separate forts.

War Department map of Fort Wadsworth, 1818/9. (National Archives and Records Administration (CC0 1.0))

Fort Wadsworth, from the Staten Island postcard collection (New York Public Library, Irma and Paul Milstein Division of United States History, Local History and Genealogy (CC0 1.0) {{PD-US-expired}})

Forts Tompkins and Wadsworth, New York (c. 1870–75), by Seth Eastman. (United States Army Center of Military History (CC0 1.0))

This is where the different names come into play. These forts were Fort Richmond, a red sandstone fort on the site that is now called Battery Weed, Fort Tompkins, Fort Morton, and Fort Hudson. The four forts had the capacity for 164 guns. The original Dutch blockhouse was incorporated into Fort Tompkins. Fort Tompkins was pentagonal with circular bastions. Fort Richmond was semicircular and had three fronts that faced the water, one front that faced the land, room for 116 cannons facing the water, and 24 howitzers on the land-side front.

Fort Richmond and the other forts were expanded during the War of 1812 with 900 cannons in situ by 1815, although none was ever fired in anger. By 1835 Fort Richmond and Fort Tompkins had structurally depreciated to the point where they were unusable. In 1847 both forts were totally reconstructed in keeping with the federal Third System of forts.

The Third System represented the acme of coastal fortifications. These massive, often three-tiered forts bristled with huge, smooth-bore cannons such as Columbiads, Rodmans, and Parrott Rifles that were the latest of their kind and most advanced in terms of metallurgy. Possessing tremendous firepower and encased in masonry and granite, these fortresses formed a protective bracelet that stretched from Maine to Key West and from New Orleans to San Francisco. Attack by a foreign naval power operating in the Atlantic Ocean, the Gulf of Mexico, and the Pacific Ocean was for all intents and purposes nullified.

New York was never attacked in the Civil War. However, there was always room for improvement. The North and South Cliff batteries were built flanking Fort Richmond that was renamed Fort Wadsworth in 1865. Two new batteries were built near Fort Tompkins. Battery Hudson included an emplacement for a type of disappearing gun, a 15-inch Rodman mounted on a King's depression carriage.

In 1885 the Board of Fortifications, also called the Endicott Board after its chairman, Secretary of War William C. Endicott, endorsed major changes yet again to coastal defenses. Part of the program included naming the whole area of installations Fort Wadsworth. Fort Richmond became known as Battery Weed after Brigadier-General Stephen H. Weed, who was killed at Gettysburg in 1863. This is a list of the batteries that were completed at Fort Wadsworth from 1896 to 1905:

- Ayres   2 x 12-inch guns with disappearing carriage
- Dix   2 x 12-inch guns with disappearing carriage
- Hudson   2 x 12-inch guns with disappearing carriage
- Richmond   2 x 12-inch guns with disappearing carriage
- Barry   2 x 10-inch guns with disappearing carriage
- Upton   2 x 10-inch guns with disappearing carriage
- Duane   5 x 8-inch guns with disappearing carriage
- Unnamed   2 x 8-inch guns on a Rodman carriage
- Mills   2 x 6-inch guns with disappearing carriage
- Barbour   2 x 6-inch Armstrong guns on a pedestal
- Barbour   2 x 4.72-inch/40 caliber Armstrong guns on a pedestal carriage
- Turnbull   6 x 3-inch guns on a pedestal
- Bacon   2 x 3-inch guns on a masking parapet
- Catlin   6 x 3-inch guns on a pedestal

Fort Tompkins panorama, in Fort Wadsworth, taken in 2016. (Ruhrfisch (CC BY-SA 4.0))

The unnamed battery of two 8-inch guns and the two sections of Battery Barbour were started shortly after the onset of the Spanish–American War in early 1898. The 6-inch and 4.72-inch Armstrong guns were bought from Great Britain.

After World War I the guns at Battery Barbour (4.7-inch and 6-inch Armstrongs) and the 3-inch guns at Battery Bacon were removed and not replaced. During World War II no new guns were added to Fort Wadsworth. From 1948–52 the fort was the headquarters of the 102nd Antiaircraft Artillery Brigade (New York National Guard), and between 1952 and 1960 it was the headquarters of the 52nd Antiaircraft Artillery Brigade until it moved to the Highlands Air Force Station. It then became the site of the United States Army Chaplain School. The United States Navy took over in 1979 and used it as Headquarters Naval Station New York. The Navy left in 1995 and all the property was transferred to the National Park Service as part of the Gateway National Recreation Area, which is what it is today.

Battery Weed, Fort Wadsworth, probably late 19th century, demonstrating some impressive cannonball stacking skills. (National Parks of New York Harbor)

Map of Fort Wadsworth, 2012.
(National Parks Service)

The fort today: The fort's address is 210 New York Avenue, Staten Island, NY 10305. The United States Army Reserve occupies several buildings on the fort, as does the United States Coast Guard and the United States Park Police. These buildings are off limits to the public. Fort Wadsworth sits on 226 acres on the northeastern shore of Staten Island. You can drive or take the S51 bus from the Staten Island Ferry Terminal to the park entrance on Bay Street.

If you're an artillery buff, Fort Wadsworth should be near the top of your bucket list of places to visit. Disappearing guns are extremely rare and there aren't many places where you can see them up close. There's also my personal favorite, the Rodman cannon, a huge, smooth-bore cannon that is often confused with a Columbiad because they look very much alike. There aren't too many places where you can see a Rodman either. The Mont Sec House is furnished as it would have been at the end of the 19th century. There are also hiking and biking trails and the views of New York Harbor and the Verrazano Bridge are something special. The New York City Marathon starts at Fort Wadsworth.

www.nps.gov/gate/learn/historyculture/fort-wadsworth.com

Fort Washington

The purpose of Fort Washington was to prevent the British from venturing up the Hudson River. It was situated on the northern end of Manhattan at the island's highest point, near the present-day neighborhood of Washington Heights. It was the only impediment that prevented the British from extending British control over New York. With its sister fort on the opposite side of the Hudson, Fort Lee in New Jersey, Fort Washington was ideally situated on a large outcrop of rock that overlooked the river. After officers Henry Knox, Nathanael Greene, William Heath, and Israel Putnam consulted with George Washington, General Rufus Putnam was given the job of building the fort. He was Israel Putnam's cousin, appointed by General Washington as Chief of Engineers of the Works of New York.

Rufus Putnam: During the French and Indian War, Putnam gained valuable experience and expertise working with the British. As a colonel, he built fortifications at Providence, Newport, and West Point. Prior to that he had worked with his Massachusetts regiment and fought the British at Roxbury, Massachusetts. In January 1783 he was commissioned a brigadier-general. When the Revolutionary War was over, he returned to Roxbury where he advocated over the years for land grants for Revolutionary War veterans and authored the Newburgh Petition that was submitted to Congress. Along with Benjamin Tupper, Samuel Holden Parsons, and Manasseh Cutler, he established the Ohio Company of Associates on March 3, 1786, the primary purpose of which was to settle lands in what was then called the Northwest Territory i.e., the land between the Appalachian Mountains and the Mississippi River that had been ceded to the United States by Great Britain under the Treaty of Paris in 1783. After the passage of the Northwest Ordinance, the company bought a million acres of land north of the Ohio River between what is now Marietta, Ohio, and Huntington, West Virginia. From 1792–3 he was a brigadier-general in Anthony Wayne's Ohio campaign that waged war against Native Americans, including the Shawnee, Lenape, and Seneca, who were ultimately defeated. In 1802 he was elected as a Washington County delegate to the Ohio Constitutional Convention. He died on May 4, 1824. His home in Marietta is a National Historic Landmark, and his home in Rutland, Massachusetts, is on the National Register of Historic Homes.

Rufus Putnam by printmakers Asher Brown Durand & Samuel William Reynolds. (Courtesy New York Public Library (CC0 1.0) {{PD-US-expired}})

Extract of map by Claude Joseph Sauthier (1736–1802) showing the chevaux de frise between Fort Lee and Fort Washington. (Courtesy of Norman B. Leventhal Map Center, Boston Public Library (CC0 1.0) {{PD-US-expired}})

When it came to building forts Putnam clearly knew what he was doing and had a long list of successes. But this time his efforts weren't going to be good enough. The first order of business was to construct a cheval-de-frise that would prevent British ships from sailing up the Hudson and outflanking the fort and Fort Lee. It took a month of grunt labor for soldiers from Pennsylvania to take large rocks from the uppermost part of Manhattan to the river's edge, where they were loaded into cribs and taken across the Hudson. Then they began work on the fort itself. An immediate problem was that there wasn't much dirt on the primarily rocky surface, so they had to lug dirt up from the ground below, but they were unable to dig ditches or trenches around the fort that enclosed three to four acres. The fort was

Claude Joseph Sauthier's map depicting the battle of Washington. (Courtesy of Boston Public Library Digital Map Collection (CC0 1.0) {{PD-US-expired}})

pentagon-shaped and had five bastions. The main walls were earth and constructed with ravelins that had openings so cannons could fire from every angle. There was an abatis around the fort.

In addition to its resident cannons and muskets, the fort had further defenses. Batteries were cited on Jeffery's Hook which stretched into the Hudson; there was another on Cox's Hill that overlooked Spuyten Duyvil Creek, and yet another at the northern end of Manhattan to control the King's Bridge and Dyckman's Bridge over the Harlem River and along Laurel Hill to the east of the fort. South of the fort were three more lines of defenses in the form of trenches and foxholes that peppered the hills. The first line was supported by a second line approximately a third of a mile to the north, and a third line a quarter of a mile north of the second. Given the number of defensive works and batteries that had to be manned, garrison strength was an inadequate 1,200 officers and men. This was arguably Putnam's masterpiece; after the barracks were completed the garrison was placed under the command of Major-General William Heath.

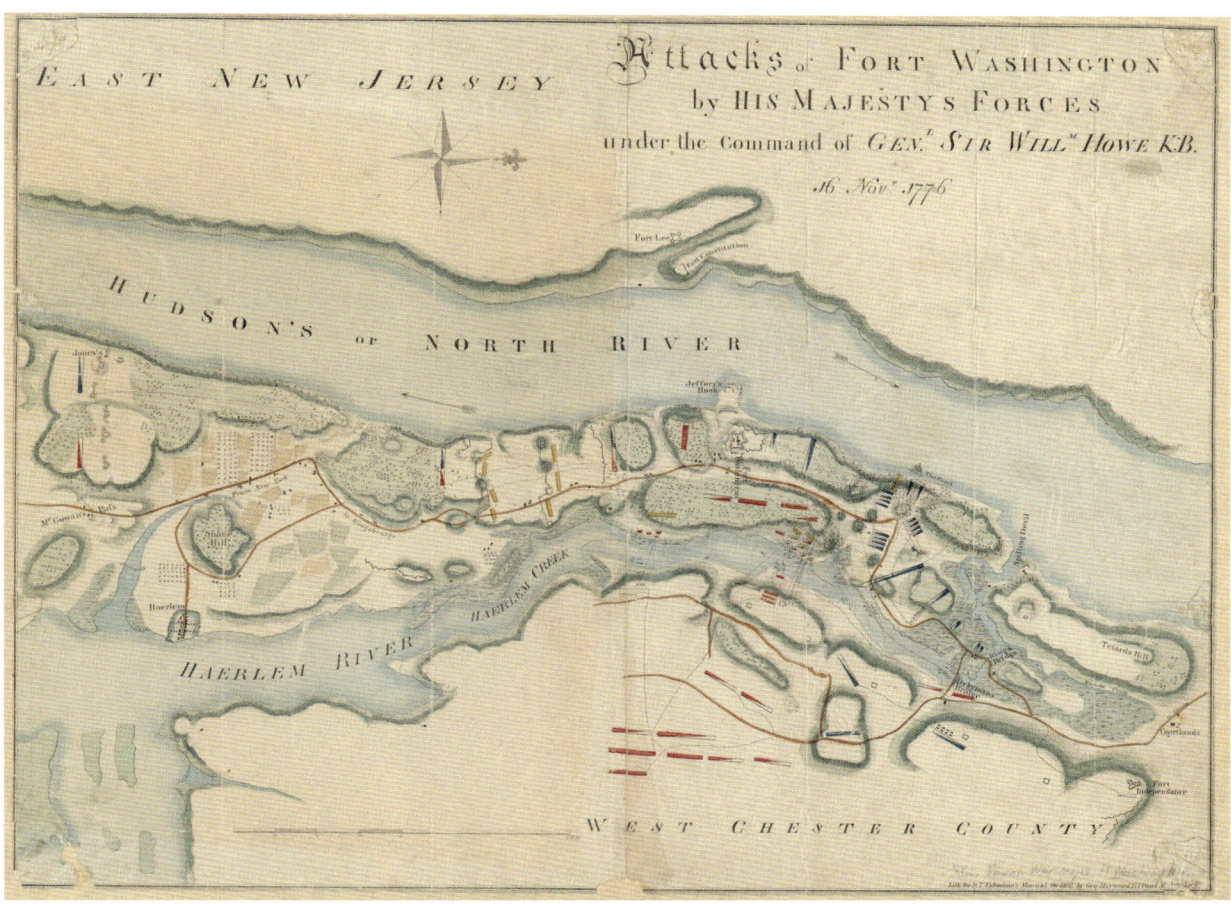

George Hayward's c. 1800 map of General William Howe's attack on Fort Washington. (Courtesy of New York Public Library Digital Collections)

William Heath was a captain in the Roxbury (Massachusetts) Company of the Suffolk County Militia regiment in 1760 and, like many of his contemporaries, rose steadily through the ranks. Prior to that, in 1765, he was elected as a member of the Ancient and Honorable Artillery Company of Massachusetts, becoming the regiment's commanding officer and a colonel. As a brigadier-general in December 1774 he was to see action at Lexington and Concord in April 1775. In January 1777 his attack on Fort Independence failed miserably, his troops were routed, his abilities questioned, he was censured, and never given command of troops in action again.

Hugh Percy, the 2nd Duke of Northumberland, was a career officer. By the age of 17 he was already a captain in the 85th Regiment of Foot. In 1764 he was made full colonel and was an aide-de-camp to the King. 1774 saw him in Boston as a colonel in the 5th Regiment of Foot that was later named the Northumberland Fusiliers. Although he suffered from gout and had poor eyesight, he was regarded as a leader who was considerate of the men who served under him. He disdained floggings and firing squads that were the norm, instead leading his men by example. He had fought in the Seven Years' War at the battles of Bergen and Minden, and in the Revolutionary War at the battles of Lexington, Concord, and Long Island.

Hugh Percy, Duke of Northumberland (1742–1817). (Courtesy of Emmet Collection of Manuscripts, New York Public Library Digital Collections (CC0 1.0))

William Dermont was born in England, and settled in Pennsylvania before the Revolutionary War. He was commissioned as an ensign in the 5th Pennsylvania Battalion on January 6, 1776, and became the regimental adjutant to Colonel Magaw. During the night of November 2/3 he snuck undetected into Percy's encampment at McGown's Pass in Manhattan, where he revealed what he knew. Subsequent to the fall of the fort, he attached himself in an unofficial capacity to General Howe's army until 1780 when he returned to England and sought a financial reward for the information he had provided. He was given a pittance.

Wilhelm Knyphausen or, more properly, Wilhelm Reichsfreiherr von Innhausen und Knyphausen, was educated in Berlin and entered military service in 1734; two years later he was a general officer in the army of Frederick the Great. By the time he arrived in the Thirteen Colonies of British North America, he was a seasoned veteran with 42 years' experience under his belt. He was immediately made second-in-command in General Heister's 12,000-strong army. He and his loyal Hessian soldiers saw action at White Plains, Brandywine, Fort Washington, Germantown, Springfield, and Monmouth. In 1779–80 he commanded New York City that was under British control. He left North America in 1782 due to poor health that included blindness in one eye caused by a cataract but died in 1800 as a result of eye surgery intended to cure his cataract and restore his sight.

Wilhelm von Knyphausen, who had the curious habit of buttering his bread with his thumb, which intrigued his colleagues no end. (Courtesy of Print Collection, Miriam and Ira D. Wallach Division of Art, Humanities and Social Sciences Library, NY (CC0 1.0) {{PD-US-expired}})

Edward Mathew entered the Coldstream Guards, also known as the 2nd Foot Guards, as an ensign in 1746 and by 1775 had risen through the ranks to colonel and aide-de-camp to George III. The following year he was in North America as a brigadier-general and commanded a brigade of guards at Kips Bay in Manhattan on September 15. During the siege of Fort Washington, he led two light infantry battalions that captured a much-needed foothold for Conwallis's troops below Laurel Hill. As a result of his actions, he was promoted to Major-General. In May 1779 Matthew, along with Admiral George Collier, made a successful raid on the Virginia coast and in 1780 he led a brigade that complemented Knyphausen's Springfield raid. He then returned to Britain later that year and subsequently became Commander-in-Chief in the West Indies.

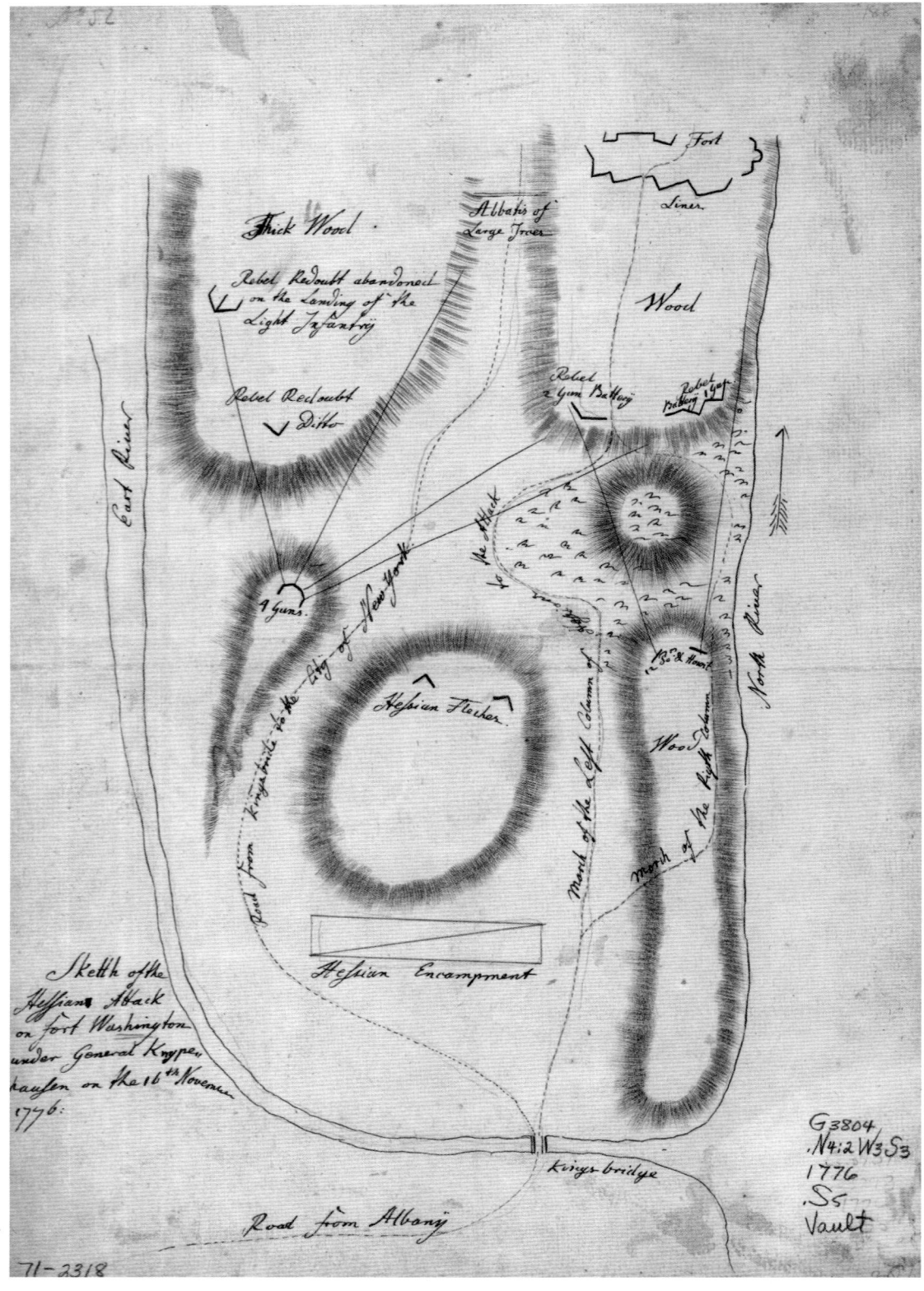

Sketch of the Hessian attack on Fort Washington under General Knyphausen on November 16, 1776. (Library of Congress (CC0 1.0))

The man in charge of Fort Washington in 1776 was Robert Magaw, who prior to the Revolutionary War had served in the militia; when the war started, he was made a colonel in the 5th Pennsylvania Battalion. By the end of August 1776 British General William Howe had gained control of western Long Island after the battle of Long Island. On September 15 he invaded Manhattan but his march north was checked the following day in the battle of Harlem Heights. The crafty Howe then landed troops in southern Westchester County (today's the Bronx) in an effort to cut off the Continental Army's line of retreat. Washington's countermove was to withdraw most of his soldiers north to White Plains. Howe then placed Hugh Percy and a small number of men below Harlem Heights to keep an eye on the fort and report back any developments that might be of tactical interest to Howe. Percy also had the authority to act autonomously.

On the morning of October 27 Colonel Magaw was alerted by watchful sentries who informed him that Percy's soldiers were initiating an attack supported by two frigates. One frigate, HMS *Roebuck*, had seen action at the battle of Long Island where it had attacked American gun batteries at Red Hook on August 27. Six weeks later, on October 9, accompanied by HMS *Phoenix* and *Tartar*, she destroyed two armed American galleys and bullied her way up the Hudson to engage Fort Lee and Fort Washington.

Contemporaneous view drawn by British officer Thomas Davies of the attack against Fort Washington on November 16, 1776; it shows artillery fire on the fort and redoubts as well as several boats of soldiers in the river. The New Jersey palisades and the Hudson River are also shown in the background. (Courtesy New York Public Library (CC0 1.0) {{PD-US-expired}})

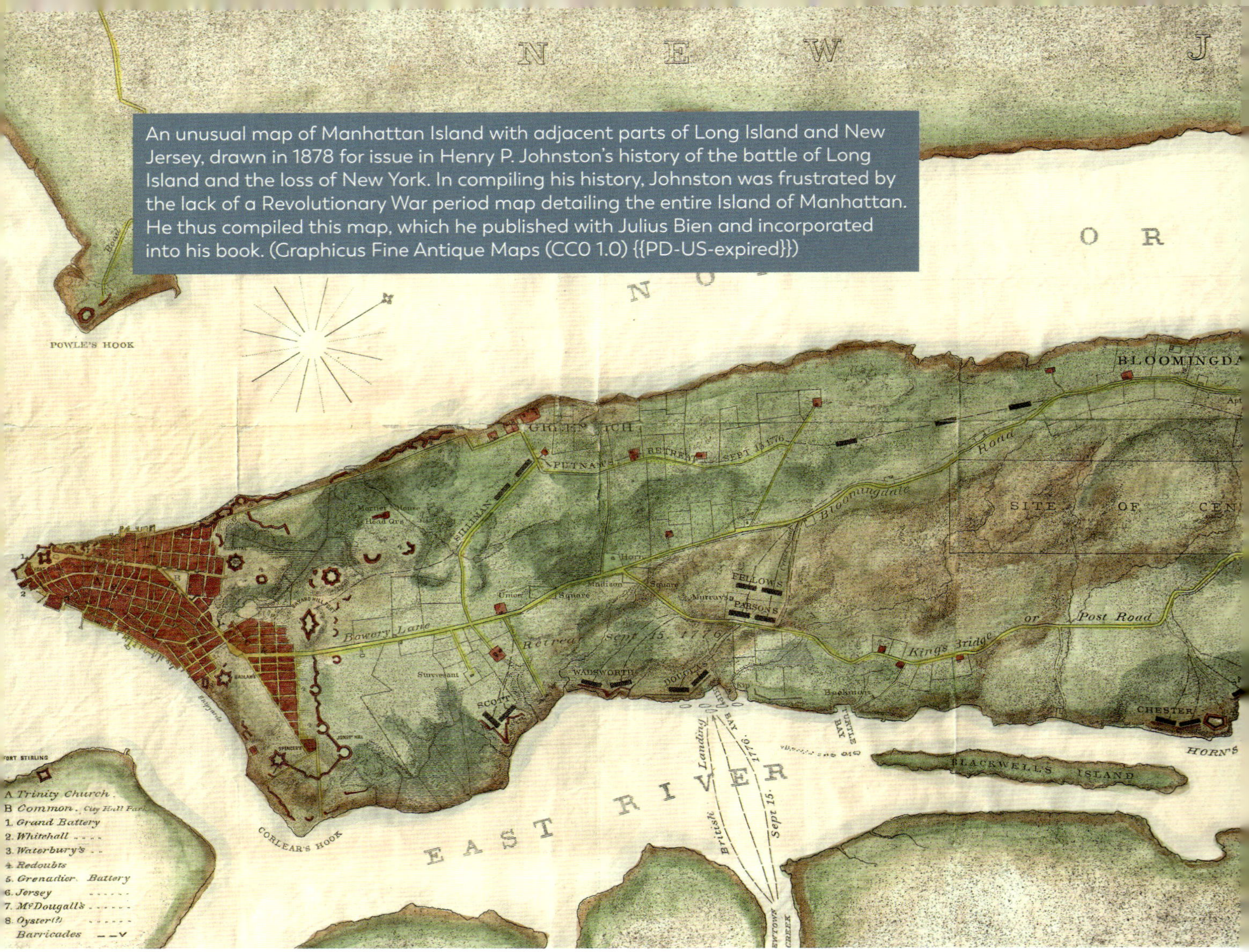

An unusual map of Manhattan Island with adjacent parts of Long Island and New Jersey, drawn in 1878 for issue in Henry P. Johnston's history of the battle of Long Island and the loss of New York. In compiling his history, Johnston was frustrated by the lack of a Revolutionary War period map detailing the entire Island of Manhattan. He thus compiled this map, which he published with Julius Bien and incorporated into his book. (Graphicus Fine Antique Maps (CC0 1.0) {{PD-US-expired}})

Johann Rall (originally Rahl). (Courtesy of Emmet Collection of Manuscripts, New York Public Library Digital Collections (CC0 1.0) {{PD-US-expired}})

The spunky American batteries immediately responded to the threat by raining down cannonballs on the ships whose own cannons couldn't be elevated to the height of the American positions. Heavily damaged, the British frigates had to be towed away to safety. On November 9 Hessian soldiers attacked the second redoubt but were driven off by 24 Americans. The Hessians were commanded by Wilhelm von Knyphausen. A week earlier, Magaw's adjutant, William Dermont, had deserted and given the British detailed plans of both Fort Lee and Fort Washington.

Percy's excursions into American lines were exploratory more than anything else. He lost a handful of men, which he could afford, while the Americans fared much better. These attacks and

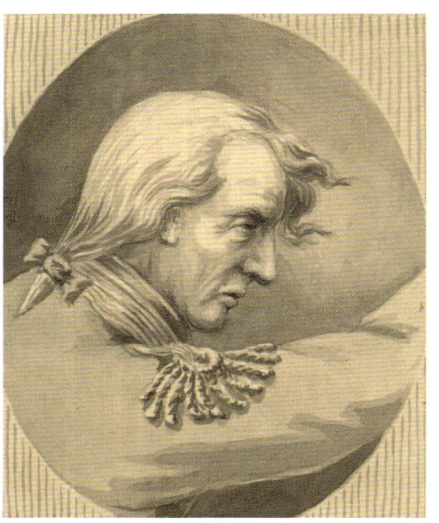

NOTE.
CAPTURE OF FORT WASHINGTON.
NOV. 16. 1776.

▬▬ Americans.
▭▭ British.

A. Col. Magaw commanding the Fort.
B. Knyphausen and Rall attacking Col. Rawling.
C. Cornwallis attacking Col. Baxter.
D. Percy attacking Col. Cadwallader.
E. Lt. Col. Stirling intercepting Cadwallader.
F. British Redoubts covering Cornwallis.

Scale of Miles.

MAP OF
NEW YORK CITY
AND OF
MANHATTAN ISLAND
WITH THE
AMERICAN DEFENCES
IN 1776.
Compiled by
HENRY P. JOHNSTON.

counterattacks took place over several days. The minor successes buoyed Magaw, who became overconfident and felt he could withstand a British siege until the end of December at the very least. Percy naturally sent the information from Dermont onto Howe, who had defeated Washington at the battle of White Plains. There was a lull as Washington and his army retreated north and the British began their assault on Fort Washington in earnest. During that time, reinforcements trickled into the fort and its garrison grew from 1,200 to 3,000.

This was Howe's plan. He would attack Fort Washington from three directions simultaneously with three forces while a fourth force feinted. Knyphausen would attack from the north. Percy would lead a Hessian brigade and several British battalions and attack from the south, and Cornwallis and the 33rd Regiment of Foot, plus General Edward Mathew and light infantry, would attack from the east. The 42nd Highlanders would land on the eastern side of Manhattan, south of the fort, which would be the feint.

Under a flag of truce, on November 15, General Howe sent Lieutenant-Colonel James Patterson with a message: if the fort did not surrender, the entire garrison would be killed. Magaw foolishly declined the offer. The American garrison consisted of the 3rd Pennsylvania Regiment, the 5th Pennsylvania Regiment, Colonel Moses Rawling's Maryland and Virginia Riflemen, and the Bucks County Militia.

The attacking British forces were a composite force of Grenadier, Light Infantry, and Foot Guards, the 4th, 10th, 15th, and 23rd Royal Welch Fusiliers, the 27th, 28th, 33rd, 42nd (Black Watch), 43rd, and 52nd Regiments of Foot, and Fraser's Highlanders. In the eerie predawn hours of November 16, Hessian and British soldiers moved out. The Hessians were ferried across the Harlem River to Manhattan but a low tide prevented the British from landing right after them, forcing the German troops to wait. At 7 a.m. Hessian cannons began to bombard the American battery on Laurel Hill, while the frigate HMS *Pearl* simultaneously bombarded the American entrenchments. Manned by a well-trained crew of 220 officers and men, the *Pearl* was a hard-hitting frigate which punched above her weight. A fifth-rated ship of the *Niger* class, her gun deck boasted an impressive array of 12 x 12-pounders, while her quarterdeck was well armed with four 6-pounders and her forecastle with two 6-pounders. Her beam was a spacious 35 feet (10.6 meters).

At the same time, Percy's artillery opened up south of the fort, aimed at the American guns that had been damaged several weeks earlier. By noon the German infantry started their methodical advance, while Mathew and his men plus Howe were ferried across the Harlem. American artillery on the Manhattan shore tried to deter them but the British charged up a hillside, scattered the American defenders, and finally reached a redoubt held by Pennsylvania Volunteers. After a brief but violent firefight, the American retreated to the relative safety of the fort.

At the same time, to the north of the fort, Johann Rall, who was leading the Hessian right, clambered up a steep hillside south of Spuyten Creek but faced no resistance. At that point the Hessians brought up their cannons as their main body, 4,000 troops, and advanced unimpeded down the Post Road that ran between Laurel Hill and the hill that Rall occupied. As they got closer to the wooded hillside near the fort, 250 riflemen from the Maryland and Virginia Rifle Regiment immediately opened fire.

Lieutenant-Colonel Moses Rawlings commanded the regiment. He had started out as a lieutenant in Captain Michael Cresap's Independent Rifle Company from Frederick County, Maryland. When Cresap was killed, Rawlings became company commander with the rank of captain. On June 17, 1776, the company was absorbed into the newly formed Maryland and Virginia Rifle Regiment and at that point he was promoted to lieutenant-colonel. The regiment had around 250 men in total and was situated about half a mile south of the fort on Manhattan Island. In a strong defensive position of rocks and fallen trees, Rawling's men somehow managed to ward off the first two Hessian charges. And then an extraordinary thing happened.

On the top of a ridge known today as Bennett Park, John Corbin was in charge of firing a small-bore cannon. Defying all acknowledged military regulations and for reasons known only to herself, Corbin was accompanied by his wife Margaret. Even a small-bore cannon made a considerable amount of noise when fired

and there was also the recoil that had to be dealt with. In any case, during a Hessian assault on her husband's position, he was killed. Margaret jumped right in and took his place at the gun until she was wounded in her jaw, chest, and left arm. (She was the first known female combatant in the Revolutionary War.)

At the same time that Margaret was firing her husband's cannon, Percy and 3,000 troops positioned to the south began advancing in two columns. The Hessians were on the left and Percy on the right. He halted after about 200 yards (182 meters) to wait for a feint orchestrated by Colonel Sterling.

Margaret Corbin by Herbert Knotel, c. 1955. (Courtesy of the West Point Museum Collection, United States Military Academy)

Facing Percy was Captain Alexander Graydon, who had commanded a company at the battles of Long Island and Harlem Heights. Magaw's new second-in-command was Lambert Cadwalader, who in January had been promoted to lieutenant-colonel in the 3rd Pennsylvania Battalion. He was to hold the three defensive lines south of the fort for as long as possible.

When it was reported to him that enemy troops had landed on the shore to his rear, he immediately dispatched 50 soldiers to thwart it. The 50 troops collided with Colonel Sterling's feint that consisted of 700 soldiers from the 42nd Foot. Hearing of this, Cadwalader immediately dispatched an additional 100 men. The British were spread out, trying to pick their way through inhospitable terrain. Some were still crossing the river. From a hilltop the Americans fired on the British, who replied by charging and dislodging the Americans. The British and Hessian assault was methodical and although American resistance was mule-stubborn, British cannon fire compelled Graydon to abandon the first defensive line and then the second, where Washington, Greene, Putnam, and Hugh Mercer were ensconced. This quartet then sailed across the Hudson River to Fort Lee on the opposite side.

Magaw understood that Cadwalader's men were about to be encircled, so he ordered them back to the fort. They were chased by Percy's soldiers who also pursued the Americans who had tried to thwart Sterling's landing and they too were forced back to the fort. Believing that some American soldiers still remained in the entrenchments, Sterling's soldiers paused, giving the retreating Americans enough time to escape.

The end was near and Fort Washington's demise was almost a foregone conclusion.

Following the complete failure of the outer lines to the south and east of the fort, there was nowhere else for the Americans to go. The fort was their only option. The fort's battery had been knocked out of action by HMS *Pearl*. Rawling's men somehow managed to hold on but barely. Overused muskets jammed or misfired.

The Hessians gradually moved up the hill, engaging the Americans in hand-to-hand combat. A bayonet charge claimed the summit.

Washington, who saw what was happening from his vantage point at Fort Lee, tried to save his men and sent Magaw a message imploring him to hold out until nightfall when the remaining men at Fort Washington could be evacuated across the river under the cover of darkness. But by the time Magaw received the message, the Hessians had taken complete control of the area that lay between the fort and the Hudson River.

Knyphausen gave the honor of requesting the American surrender to Johann Rall, who in turn sent a certain Captain Hohenstein, who was fluent in both English and French, under a flag of truce. Captain Hohenstein then met with Cadwalader, who requested that Magaw have four hours to consult with his officers. Hohenstein gave him 30 minutes. To his credit, Magaw did everything he could for

Colonel Lambert Cadwalader (1743–1823), an oil painting by Charles Willson Peale, 1771. (Philadelphia Museum of Art (CC0 1.0) {{PD-US-expired}})

American militiamen at Fort Lee (reenactors), 2017. (Jim Henderson (CC BY-SA 4.0))

his remaining men but in the end, they were allowed to keep only their personal possessions. The decision to formally capitulate was made at 3 p.m. and at 4 p.m. the American flag was lowered and replaced by a British one. When the Hessians entered the fort, they immediately began looting. A total of 34 cannons along with two howitzers, ammunition, tents, blankets, and anything else of value was seized.

Washington's messenger, Captain John Gooch, had arrived at the fort just before it was completely surrounded. Before the formal surrender, Gooch, desperate not to be taken prisoner, leapt off the side of the fort, survived a rough head-over-heels tumble down the bottom of the cliff, somehow managed to avoid being shot or bayoneted, found a discarded boat, and returned without serious injury to Fort Lee.

Reenactors at the Fort Lee blockhouse. (Jim Henderson (CC BY-SA 4.0)

Surrender of Rall's Hessians to General Washington after the battle of Trenton. (National Archives at College Park (NARA) (CC0 1.0))

Most Americans were not as fortunate: 59 were dead, 96 were wounded, and 2,873 had been captured. After they were paraded through the streets of New York City, the prisoners were interned on British ships in the harbor. Over 2,000 of them subsequently died either from cold, disease, or starvation in the bitter winter. The rest were released 18 months later in a prisoner exchange. Magaw fared better. He was paroled and confined to New York City, where he met and married Marritje Van Brunt in April 1779. She bore him two healthy children. By way of contrast, the British only lost 84 men killed and 374 wounded.

Fort Lee was abandoned three days after Fort Washington fell. The Continental Army was in headlong retreat through New Jersey and across the Delaware River into Pennsylvania northwest of Trenton. The British pursued them as far as New Brunswick.

The fort today: The site where Fort Washington once stood is now Fort Washington Park. It is 160 acres and bordered on the north by Dyckman Street, on the east by the Henry Hudson Parkway, and on the south by 155th Street. Stones mark the location of Fort Washington's walls and there is a tablet that informs readers that the outcropping is the highest point on Manhattan Island. New York's last remaining lighthouse, "the little red lighthouse," is located on the Hudson River under the George Washington Bridge on the Manhattan side. The lighthouse can be seen in several scenes from the 1948 film noir *Force of Evil* that starred John Garfield, Marie Windsor, and Thomas Gomez.

A month later, on the night of December 26/27, 1776, Washington crossed the Delaware and defeated a Hessian garrison under the command of Rall at Trenton. Washington's next victory was at Princeton, New Jersey. It took seven long years before the American flag was raised once again over Fort Washington, on November 25, 1783, as the last British forces were leaving New York and heading home. The saga of Fort Washington had finally come full circle.

Conclusion

The slow burning fuse on the powder keg that was to become the American Revolution reached its acme in April 1775 when General Thomas Gage and a large number of British regular troops totaling four regiments, or approximately 4,000 soldiers, attempted to apprehend the colonial leaders Sam Adams and John Hancock in Lexington. From there Gage intended to go to Concord, where the two colonial leaders had stashed a considerable amount of gunpowder. Alerted by the renowned silversmith Paul Revere who rode to Concord, 20 miles (32 kilometers) northwest of Boston, the colonists were waiting for the British.

Stand Your Ground by Don Troiani. (Courtesy of the National Guard)

Greased & Hellyards cut away

Fort George. (New York.) bearing ESE Distᵗ ½ Miles, at the Evacuation

A panoramic view of Fort George, New York, on the day of the British evacuation, November 24, 1783, a watercolour by Warrant Officer Robert Raymond RN. (Courtesy of www.christies.com/ (CC0 1.0))

Gage and 700 British troops mustered on Boston Common on the evening of April 18, 1775. They crossed the Charles River at 2 a.m. The march took them through a waist-deep swamp and when they finally arrived at their destination, it was 5 a.m. On the Lexington Common, 77 armed colonists opened fire on the British. At the end of the fight, 95 colonists were dead or wounded. The British lost a total of 273 killed or wounded and 26 were deemed missing. Gage retreated back to Boston. The powder keg had been lit. America's Revolutionary War had begun.

On September 28, 1781, 16,000 American and French soldiers, led by General George Washington and the French commander Comte de Rochambeau, laid siege to the Virginian town of Yorktown, held by British General Charles Cornwallis. Utilizing Vauban's technique, it took the Americans and French about three weeks before the British finally surrendered. The defeat at Yorktown ended for all intents and purposes the Revolutionary War. Claiming illness, Cornwallis sent Brigadier-General Charles O'Hara in his place, while Washington sent his second-in-command, Benjamin Lincoln, to formally accept Cornwallis's sword. It would be more than two years later, on November 23, 1783, that the British finally evacuated New York, opening the way for Washington and the Continental Army to return in triumph.

In the intervening period, from Lexington to Yorktown, a span of over six years covering a panoramic canvas where a vast cast of characters—British, American, French, Hessian, Native American, and civilian, heroes and villains, traitors and spies—fought and died, the American Revolutionary War played out. In the array of actions that took place, from reconnaissance missions to ambushes, raids and massacres, sieges and assaults, to booming naval bombardments, inevitably the common denominator was the fort. From the humble fortified stone house with musket loopholes to the imposing star fort with bastions and ravelins that boasted dozens of cannons and hundreds of defenders, the fort was often the nexus of a raid, a battle, or a campaign. Many were temporarily thrown together in a matter of days or weeks and deteriorated as quickly when they were no longer required. Many made way for development. Others, grander and more enduring, have survived to this day and are protected as a reminder of that epic conflict that echoed around the world some 250 years ago.

Bibliography

American Battlefield Trust. *Battle Maps of the American Revolution Volume 3*. Knox Press: Louisville, KY, 2022.

Bicheno, Hugh with Holmes, Richard. *Rebels and Redcoats: The American Revolutionary War*. William Collins: London, UK, 2014.

Black, Jeremy. *The War for American Independence, 1775–1783*. The History Press: Cheltenham, UK: 2021.

Chartrand, René & Spedaliere, Donato (Illustrator). *Forts of the American Revolution 1775-83: 110 (Fortress)*. Osprey: Oxford, 2016.

Fisher, Sydney George. *The True History of The American Revolution*. Palala Press: 2016.

Harris, Michael C. *The Philadelphia Campaign, 1777*. Casemate: 2023.

Harvey, Robert. *A Few Bloody Noses: The American War of Independence*. Robinson: London, UK, 2013.

Hourly History. *American Revolutionary War: American Revolution, Boston Massacre, Boston Tea Party, Battles of Lexington and Concord, Battle of Bunker Hill*. Independently published: 2021.

O'Neill, Bill & Walker, Dwayne. *The American Revolutionary War Trivia Book: Interesting Revolutionary War Stories You Didn't Know: Volume 5*. CreateSpace: 2018.

Ronald, D. A. B. *The Life of John André: The Redcoat Who Turned Benedict Arnold*. Casemate Publishers: Philadelphia, PA, 2019.

Smith, Digby & Kiley, Kevin F. *An Illustrated Encyclopedia of Uniforms of the American War of Independence 1775–1783*. Lorenz Books: Dayton, OH, 2008.